INDUSTRIAL STATISTICS

Professional Journey

by Babu

DORRANCE
PUBLISHING CO
EST. 1920
PITTSBURGH, PENNSYLVANIA 15238

Dorrance Publishing Co
585 Alpha Drive
Pittsburgh, PA 15238
Visit our website at www.dorrancebookstore.com

ISBN: 979-8-88812-996-8
eISBN: 979-8-8881-2570-0

INTRODUCTION

I dedicate this book to my loving wife Alice Matthew, who supported me throughout my life, in good times and bad, and to my beloved daughters, Reetha Matthew, Anna Matthew and Caroline Matthew.

John: 15:16

"It was not you chose me, says the Lord, but I who chose you and appointed you to go and bear fruit, fruit that will last".

Luke: 8: 16

"No one who lights a lamp conceals it with a vessel or sets it under a bed; rather, he places it on a lamp stand so that those who enter may see the light".

Matthew: 5: 14-16

"You are thee light of the world. A city set on a mountain cannot be hidden. Nor do they light a lamp and then put it under a bushel basket; it is set on a lamp stand, where it gives light to all in the house. Just so, your light must shine before others, that they may see your good deeds and glorify you r heavenly Father".

"Scripture says (Matthew 13:8) But the seed fell on rich soil, and produced fruit, a hundred or sixty or thirty-fold."

It was not my desire to write a book, but the circumstances made me think that I should write a book that goes in detail the work I did throughout my professional career.

I am an engineering graduate (mechanical/automotive engineering) from Kerala University, Kerala, India in 1967. After my graduation, I had to wait to find a job according to my education, it was not that easy to get a job right after the graduation, and I applied for a job many places. One of them was to the ministry of defense tank manufacturing factory in Avadi, Madras. Since the waiting time for a result was uncertain, friends and I decided to start tutoring students from eighth grade to tenth grade that is high school, at that time. A couple of families near my place accepted our decision and gave their children and one family offered their vacant house to use as a place to gather and teach the classes. Initially, we stated with math and science, English and Hindi. The textbooks we selected were the same as they taught in the school at the time.

A month after our starting the school, we got overwhelming requests and support from the families around to take their children from sixth grade and up. It was a big challenge for me to schedule the classes early in the morning and complete tutoring before it was time to go to school. Our tutoring location was just opposite the school nearby and it was very convenient for the students to get back to their classes on time.

We got overwhelming approval from the parents and we decided to continue the tutorials until I found a job. In the meatime, I got an call from the government of India, ministry of defense, to attend an interview in Maharashtra, North India and the government furnished travel arrangements and accommodation. The interview was a combination of a written test and a panel interview, an entire day process. I scored 100% in the written test and the next step was the panel interview. The panel consisted of an army officer colonel, scientists from different areas of manufacturing sectors, metallurgists, and human resources. The Interview was a lengthy one and I performed well. They informed they will send a letter of appointment at a later date, and I returned home and continued the tutoring.

After eight months passed, I got an appointment letter from the government as a technical supervisor and the location was Ahmednagar, near Pune, in the October 1968. After two years of service, I took the departmental promotion test and based on the first rank, I got transferred to the Madras office and after three weeks, my next move was to Hyderabad, India, where the job assignment was to manage and control the suppliers who manufacture and supply defense related armored vehicles, and other products. After two years, I was transferred was to the Delhi, India zonal office where my assignment was supplier management of various manufacturers who were the contractors from government. The function of my department was the indigenization of assemblies and components for tracked vehicles such as the tanks, heavy earthmovers, komatsu, international harvesters and validation and approval of newly designed products for defense tracked vehicles, etc. One of my suppliers had a contract for making rubber O-rings for various tracked vehicles of defense. As a qualification of the product, the supplier must submit parts along with material certificates both physical and chemical. This supplier had submitted the parts along with material certs, but physical examination of the O-rings failed because of the trapped air, so the O-rings were spongy at certain points and the lot was rejected because of this failure. The supplier submitted the second batch with the samples of O-rings and material certificates. The second lot was also rejected due to the sponginess condition. The supplier had only one more chance to submit the samples with material certificates. The supplier was a small business owner who started the business by taking loans from the bank and that is his only income from the business. The shop owner requested my help to resolve the issues and I refused to help him because I was not a rubber technologist. The owner holds a Master's in rubber technology. But because of his persistent request, I promised to visit his shop to evaluate the manufacturing process of the O-rings.

THIS IS MY PROFESSIONAL JOURNEY TO THE PROCESS CONTROL, QUALITY, AND SUPPLIER MANAGEMENT.

[NEVER USED SPC IN ANY OF THE PROBLEMS RESOLVED]

Chapter One

My first problem solving involvement at a supplier in India

Please note that I have never used the statistical process control to manage issues related to quality and manufacturing processes and product related issues. I have used my technical skills to identify the problems you will read in this book, and I have succeeded in resolving issues and delivered quality products to our customers.

This supplier had a several million rupees contract to make rubber O-rings for the international harvester. That means all kinds of O-ring parts for the maintenance of the vehicle.

According to the contract, this supplier was supposed to submit the first article inspection/test report of the raw material, rubber, and a dimensional report of the O-rings. The first attempt failed in the final product itself. There were blisters in the O-ring, but the material compound and physical and chemical analysis were found acceptable. According to the findings, the pilot first submission of samples was rejected and asked the supplier to re-submit next lot for approval. It is particularly important to know that the supplier has only three chances to for the submission of acceptable samples and if all the three samples fail, the whole contract would be canceled and the supplier will lose the entire investment he made on this project.

A second sample was submitted with the material test data, dimensional test report, and the samples of O-rings for review and approval. This time everything met the specifications except the O rings exhibited flaws internally and the lot was rejected.

This was one of the critical moments of this small company and the owner was desperate to see what went wrong and how to solve the problem. He confessed that he sold all the gold his wife had and took some loans from the bank to get this business started from the ground up. He is a rubber technologist, he did everything to his ability, and he was saddened to see the samples submitted were rejected. He asked for my help in this situation to help him to identify the problem.I told him that I am not a rubber technologist and I have no idea how to help you out. He was in tears and I could see his miserable condition. If he loses this contract, he will lose all the investments he made and the family will go bankrupt. Because of his persistence, I decided to visit his manufacturing plant to see the manufacturing process. This small shop was laid out professionally. It was a clean shop environment, the tools and equipment were in good condition, the calibration system met the requirements, the operation instructions were good, temperature and pressure control of the press were in good and calibrated.

He showed me how the rubber compound was mixed and made into sheets before the sheet put on the hydraulic press. I asked the owner to demonstrate how he does make the O-rings using this hydraulically operated press. He started with loading the rubber sheet in the bottom of the mold and he lowered the upper arm of the hydraulic press, applied heat and pressure, and set the timer to cure the rubber.

Upon reviewing the operation of the hydraulic press, one thing struck my mind and was brought to my memory in my mechanical engineering mind which highlighted a likely cause of failure of the O-rings. The load applied to the hydraulic press was not uniformly distributing uniformly across the whole area of the hydraulic press. When I saw this, I shared my findings with the owner and suggested he should modify the press by welding steel bars/plates across from the center of the middle shaft to the diagonally opposite of the rectangular shape of the press. He agreed to do so.

After two days, he produced the sample parts of O-rings with the material test report along with the samples of O -ings. Upon evaluation of the O-rings, they found them to meet the requirements and accepted the manufacturing process. According to the procedure, those samples were sent for fitment trial on the vehicles for performance evaluation. Finally, the fitment test report was received. It was good, and the final approval was given with clearance to go ahead with the production of the O-rings. The owner was so happy that my visit to his shop was very fruitful, and I had identified the problem with his hydraulic press.

This is the picture of the press before and after the modification.

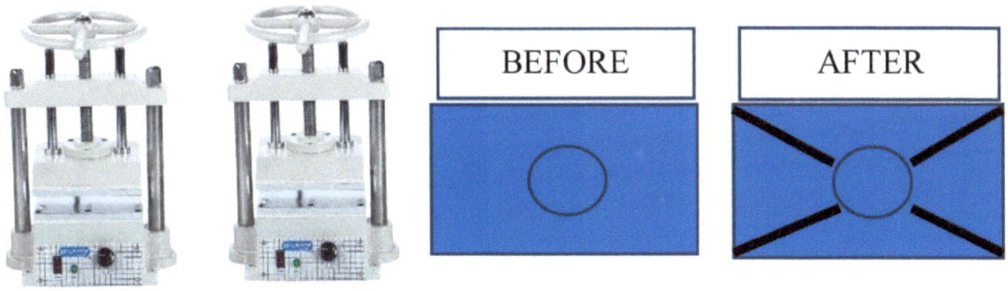

Front elevation of the mold and modified upper plate before and after.

Chapter Two

Schultes Experiment, Inc.
(How the small machine shop turned into a high yielding production shop.)
Quality Manager/Chief Inspector/Production Control and Customer Liaison

I was looking for a job in Chicago, Morton Grove, Illinois, around 1976/1977 and I noticed a hiring notice placed on the lawn of a machine shop. They were looking for machinist and I decided to go inside and ask about the position as a machinist that they posted outside. Upon inquiry they said they are looking for an inspector. Then I told them that I would like to apply for that position and they gave me an application form to fill out. After completing the form, I gave that to the office and few minutes later one gentleman came out and asked my experience with reading the engineering drawings and using the precision measuring instruments and I said, "Yes, I can read the drawings and know how to use the precision measuring instruments."

He told me to stay back, and he brought a vernier caliper and asked me to measure a round machined piece. The beauty of the vernier caliper was it is engraved in fractions. I never saw or used such a fractional caliper. I sat there for a moment and recollected my memory of how to read a regular caliper and the theory in the main scale and the vernier and I was able to measure that work piece and told the person the results of my reading. He went inside and came out with the good news that I was selected for the inspector position and asked me to come to work the following week.

The following week, on Monday, I reported to work and the gentleman who interviewed me took me to the shop and briefly explained what my duties are and showed me the inspection room that is my office. I did not know where to start or how to start my day and decided to go around the shop and meet people and introduce myself and my job there.

I had several challenges in front of me and the main issue was communication. I knew English, but the way I spoke, no one could understand what I was saying and I had to repeat it several times to make the other person to understand me. That was the style of my handling the English language in India like flowing water. That did not work in USA. So, I decided to slow down the way I communicated and gradually I gained control of my vocal communication style, and everyone was happy.

The next enemy was the working relationship with the employees and the culture of the company. Each machinist in that shop was working for that company as an independent machinist. He gets orders from the owner of the shop with the materials for that job and the machinist takes charge of the order. He makes the part according to the engineering drawing. He cleans the part or parts,getst the shipping documents ready, and ships the order to the customer. That was their normal operation's system. After I was introduced as their inspector, each machinist brings the part with the drawing to my office and tells me this is the part for inspection and I need it right now. Then I tell the machinist to give me some time to read and understand the drawing and make the measurements to verify the conformity to the requirements and they walk away. During my inspection of the part, they come several times and asked me. It is not done yet and I cooperate with the machinist and identify the good work or the things they did wrong and tell them how to correct it with my little knowledge in the making of parts.

Slowly and steadily, I captured the mind of all the machinists in the shop and they realized that this little Indian is smart after all and he is a helping hand to do their job right. My daily assignments are gradually changed to plan proactively the upcoming job orders and get the gauges, tools, and instructions ready before the job starts and issue the directions to each machinist and monitor their work routinely. I spent some time with each employee in the morning and chat with them to get their presence of mind on the job and make sure of what they are making and check for the proper use of the gauges and tools to use them properly to measure the quality of the job. I kept my office open for any question they might have during the machining of the job they have given.

My next step was to get the management/owners' attention to bring some discipline in the shop and I requested the owner to call for a meeting during the coffee break for five minutes and tell the employees to go through the quality department to get the first part approval before commencing the production. The management did exactly what I instructed. Gradually, I noticed that the performance of the shop increased, the production rate increased and none of the shipments were rejected by the customer. We started getting more orders and the shop grew a status as one of the best high precision machine shops in that region and the customers that previously condemned this shop started coming back.

It is better to look back at the history of this shop. This shop was owned and operated by immigrants from Germany and the owner worked in one of the companies in that region. He started buying old machines like drill presses, lathes, bridge port mills, surface grinders etc. He filled their garage and gradually opened as a machine shop making development parts in small quantities and whatever profit he gets was good for him because all the machines were paid off, only the employees' wages and other things were the expenses, that is all. So all the machines in that shop were used and old machines.

My first ACROLOC 12 tool holder.

We were in a good position, getting more orders and we used to make and deliver quality products and the number of customers increased. We had a job with extremely high precision, a complicated configuration made of bronze, when we ordered our first Acroloc 12 tool holder mill. After installation of the machine, the installer tried to program the machine and tried one of the bronze parts to mill a bore with 0.0001" tolerance. The first part was rejected because the bore was in taper. After a few tries, I was unhappy with the machine and the manufacturer of the machine sent some officials from their New York office. He verified the program and made some corrections to the program (All the programs were on tape with many holes punched) and then he ran the job of milling the bronze part. After inspection I noticed improvement, but it was not exactly correct. The bore was in taper condition. As a final attempt, they tried one more part and amazingly the part came out perfectly. I asked them to repeat the work a few more times and all the parts came out perfect to the drawing requirements. I was happy and I told my manager/owner to go ahead to sign the contract to buy the machine to train our machinist on.

Gradually, our shop was getting more machines, replacing the old lathes, mills with new Harding lathes, mazak mill and three more Acroloc 12 tool holder mills to run a long-rung job for making dental grinding and cleaning handheld equipment, designed by American Dental Corporation, that was our main challenging job of our creativity. We got the new engineering drawing from American Dental Corporation to make this assemble for thousands of pieces. My manager/owner hired an engineer to manage this project. I could see this engineer making a fixture of his own design and the mill fixture we were making in our own tool shop and tool room in charge was Frank. I asked Frank what was going on there and what he was making, and he told me that Gene wanted me to make things as per his direction and there were no blueprints or sketches. Gene never communicated with me at all, and I knew for sure that the first product must go through me to get approved before production begins.

After a few weeks, the milling fixture was ready and mounted on one of the Acroloc mill. Gene gave the operator instructions how to load the parts and take them out after completion of the operation. I got five parts for approval, and I checked each part and documented the results and found that each five of the parts were in five different configurations and rejected the first lot. That process continued for several days; they could not make even one batch of good parts.

The senior owner Aarun, father of the owner (son), used to visit the shop in the morning and come to my office to have coffee and chat with me for some time, then I would walk along with him in the shop to explain him what was going on in the shop, the critical parts in process etc,. This dental grinding machine part has the main body made from 303 stainless steels, and we have the stock ordered to 0.611 +/- 0.0001" ground, that will get turned on the Harding lathe to cut thread and an off-center tail with a hole drilled through. That part would go into the milling fixture that Gene made.

One day I took the senior owner to the shop and showed him the stack of rejected parts. I told him that the problem was the mill fixture and the lack of communication between Gene, me, and the tool maker Frank.

Arun grabbed a few of these rejected parts and went to the son (OJ) and asked him, "How come you didn't communicate with Bob during the design of the mill fixture? He has better ideas and some like that in their family coffee break."

So, OJ came running to my office and told me, "Bob give me the design of the fixture you have in mind."

And I said, "Please send Gene, the engineer hired for this project, to me and I will tell him what to do and how to design a new fixture to hold the blanks for machining using the CNC machine."

Gene came to my office and asked me for the design of the fixture, I gave him the sketch of the mill fixture and told him that our tool man would make the fixture complete here in our shop and send them out for heat treat,the rest will be done in our shop. Our tool maker Frank started working on the fixture according to my instructions. When completed, he sent them for heat treat, when we got them back, we did little final honing of the precision hole 0.611" Dia. and completed the assembly. After my inspection the mill fixture was given to Gene to try on the machine. This new fixture has a unique advantage is that it can hold five parts in one set and the fixture can be turned ninety-degrees to machine the other side of the part without removing it from the fixture. After the completion of machining, the part is finished, only to remove burrs from outside of the part. The deburring process was damaging the appearance of the part and I had developed a process change in the machining operation itself, since the process control was tightly controlled from the beginning, I had the confidence to make slight change that would eliminate the deburring process by about 95%. So, I asked the CNC programmer to raise the whole bed of the CNC mill about 0.005" or make the spotting tool go 0.005" deeper, whichever is easier to do, and the programmer did so.

Amazing to say that the beauty of the finished product was excellent. One fixture would take five parts for milling, and it would take almost an hour to machine completely and the first batch of the parts were brough to my office for inspection. After inspection, I noticed that all the parts were alike with no dimensional variation at all. I asked them to make a few more batches to ensure the consistency and repeatability of the process, and the final approval for production was given to make the parts. After successfully launching this production program, we ordered two more Acroloc 12 tool holder milling machines and we made the same design of two fixtures for each machine. We were making parts like popcorn. We were making several critical parts, short run and long run orders. We never did SPC to control the process, process control was done during the early part of the machining process planning that include the fixturing and measuring techniques and tools to use and operator training for monitoring the process and measure the parts at certain intervals. If any dimensional variations were noticed outside the requirement, then the operator stops the machine and reports the matter to QC department (quality control department) for investigation or further action.

Within two years, that company grew to a multimillion-dollar business, a high precision machine shop. We started a second shift; our productivity increased. We got rid of all the Second World War-type old machines and installed new automatic lathes, mills, and Harding lathes.

When I joined this company, my drive was to join the ASQC and learn the technique to learn and use statistical quality control and statistical process control. And I contacted ASQC office and paid my membership dues and got my membership in 1977. That is my best guess, and the membership number was 00039681. After my membership, I took numerous classes and training on statistical process control and statistical quality control and later I got my green belt certification from the Villanova University. Later, I came to know that the SPC and SQC, Six Sigma, Lean Six Sigma etc. are nothing but a fake education propaganda to entice the people sucked into the program by showing huge salary benefits after taking these courses and getting certified. It may be a useful tool and technique in some other areas of business but it is no good in the manufacturing industries. During my stay at the machine shop (Schultes experiment) I tried to implement SPC, but it did not work, especially when you are dealing with high precision one part or fewer parts where customer demands 100% quality, SPC will never work.

Let me tell you something about the origin of SPC, this is my best understanding. During the World War Two or after the war, the ammunition made at different parts of the United States of America did not fit into guns and that made a big impact on the war. At that time, the process was controlled by the operator of the machine itself, a skilled operator would make good parts and the less experienced machinist or the operator with less experience would make unacceptable parts. When the Industrial Revolution started, controlling the quality of the parts was a big headache and at that time the sampling inspection was introduced. From the manufactured lot, randomly select parts and measure them, document the results, and decide whether to accept the lot or reject the lot.

The manufacturing industries have undergone many significant changes since Dr. Shewhart first developed control charts back in 1920s. That was the beginning of the introduction SPC or statistical process control and from there started many different forms of statistical applications to determine the quality of the manufactured products or assemblies. Just think about this, this technology or the system started in 1920s I guess, when the machining or manufacturing technology or the process control was in a primitive stage. All the manufacturing machines were manually controlled and operated by humans, so, the skill and knowledge of the operator was extremely critical in making good parts. It amazes me that till today the business organizations like ASQ, SME, universities, colleges and consultants use the same old technique to lure or entice the community or the business organizations by publicizing the benefits and huge salary benefits for those who take the classes on SPC, SQC, Six Sigma, Black Belt, and many more branches of statistical theory and its applications. But they never told the candidates the applicability of this tool in what industries more specifically. To me it is fake or a scam, just cheating the community and misguiding them with wrong information. Poor people who are looking for a well-paying job will fall for this kind of education or training program. I can prove that these techniques are not good for the manufacturing industries, and they teach with no objectivity. The Cp, CpK, Six Sigma and many things they teach are nothing but nonsense and have no merit in the manufacturing industries.

I developed techniques during my work at the Schultes experiment, to handle the challenging quality requirement and compliance to the requirement were the prime objective and there was no room for taking measurements, plotting charts and graphs to prove the capability of the process to prove that whether the parts were good or bad. Our company policy was to make 100% quality products regardless of the quantity produced. With that in mind, I developed and created a system that I call now is the reverse statistical process control (R SPC) and now I have a software program that will take you through the basics of process control elements from the beginning till the end. This software could plot X-bar R-charts and the frequency distribution charts to use as a visual tool for the management to see the quality of the products manufactured. Please visit www.qem-inc.com for more details and the training needs of your employees. In a later part of this book, I will explain the application of traditional SPC and the application of R SPC, a proactive approach to build quality into the process.

What you take home from this example?

Quality is not something to be dictated through procedures and quoting words from the ISO or Q9858. That creates fear and frustration to get the job done. Fear always makes errors on the work they have on hand. Instead, create communication that's simple and easy to follow, practicable, and trains them to take the fear out. You will be amazed to see the productivity go up and the rejection go way down. Quality is not to be dictated but it must be planted or tailored into the process. You, the quality person, should have profound knowledge of the process. Without that, it is impossible to plant and grow quality in the manufacturing field. Statistics has no room on the manufacturing floor, that I have demonstrated through my experience and guided the employees in the right direction and provided the tools and techniques to verify their work. Statistics is a numbers game. Elements of statistics cannot be implemented without any data, that is a technology. Learn those techniques and you can sow the quality seeds right from the beginning of the manufacturing process. Those who are interested in to learn more about the technology, please contact me for a training class set up at your facility.

Chapter Three

Colt Industries, Menasco Inc., Burbank, California
MANUFACTURER OF LANDING GEAR ASSEMBLY.
Quality Assurance Engineer.

Colt Industries is the manufacturer of main landing gears for the major aerospace industries in the US. Mr. Bob Maguire, quality manager, hired me as quality assurance engineer with the responsibility to manage all the suppliers of raw forgings and castings.

This is the point I wanted to discuss here. One of the suppliers, they make titanium forging for the landing gear had a history of high rejection rates, during machining of the forging, a manufacturing engineer must stand by the machinist to change the set up to get at least sixty percent good, machined parts and the rest were rejected and scrapped. This was an incredibly challenging task for me, and I contacted the forging company and made an appointment to visit their facility to monitor the forging process, measuring technique, the quality control procedure, etc. Before my visit, I told the QC manager of that forging company to set up one forging for layout inspection and leave the set up for evaluation. When I visited there, the QC manager took me around the shop and explained various processes of forging process and the quality control system and methods, qualification of the forgings, etc. Then we came to the inspection department where the forging was set up for lay out inspection and all the lay out operations have completed;

The inspector said, "Everything looks OK."

I asked the inspector some of the missing lay out portion to verify the concentricity of the forging between the inner circle to the outside diameter, that portion he purposely eliminated thinking that I may not detect what he omitted on purpose, that was the area giving trouble, the trouble was the concentricity of the ID vs OD was off by about 0.50". From the symptom of failure, I knew the guide pins might have worn out, causing the shift of the ID and OD concentricity. I reported this finding to my manager and issued a corrective action notice to fix the problem. After a week, the supplier offered another forging for inspection. Upon receipt of the request, I instructed them to lay out and do the inspection and leave the set up for evaluation. I noticed that the supplier did not make the correction to fix the problem. I reported this matter to my manager and from his direction, I issued cancellation of the contract due to noncompliance to fix the problem and a copy to the contract department. Soon after release of the cancellation of the contract, the supplier fixed the problem, and the first shot of the forgings was offered for inspection. Upon review of the forging, they immediately fixed the guide pins of the forging die, and the next shot of the forging came out very good and the problem went away.

In this problem-solving process or identifying the cause of failure, first there was no data. Data of what? The only data we had was the rejection rate, from the failure mode, that identified the shift of the wear and tear on the guide pins of the forging die and that was confirmed on the on-sight review of the die and the part they presented for inspection. Remember that I never used SPC to resolve the problem. Issues like this are not kind of relaxing and collecting data and a witch hunt to identify the problem. When immediate actions are required, SPC process would not help, but the technical knowledge will guide you to get to the point to identify the problem. Again, it has been proven that SPC is not a good candidate in resolving issues like engineering matters.

Chapter Four

Mcdonnell Douglas Astronautics Company, Monrovia, California.
Quality Assurance Engineer.

I was hired as quality assurance engineer on a new project MMS (mast mounted sight). This is an electronic night vision camera capable of detecting objects hidden on the ground regardless of the weather condition, rain, fog, dust, smoke, or any other media, and transmit the information to the commanding center for necessary action.

This project designed by the Huntington office and the Monrovia plant was chosen to fabricate the whole assembly.

My job assignment was total quality assurance of the project, including the in house manufacturing process control (review and approve manufacturing planning, tooling, first article inspection, and approval of the machining process), supplier management, outside supplier control and employee training.

One of the key components in the project was a 12" X 12" X 12" cube, made of 99.9% beryllium copper (a very toxic material). Before machining the part, it would weigh around sixty pounds and when it gets machined, the total weight of the part would about two pounds and the profile of surface was overall flatness in three axes were 0.00005". The supplier who machined this part was in North Carolina, I believe, and the machining center was three stories-below the ground level. The area was completely airtight, and the machine operator was in a space suite. They have a positive chip evacuation process to suck out all the particles disposed in a different area because of the toxicity of the material. After machining, this part will go through a wipe test and measure the particles to a certain level and the supplier must furnish all the test data, including the measurements and material certificates.

I created a special handling procedure of this part upon arrival from the supplier because it was very fragile and toxic. I created special inspection instructions for the receiving of the housing, created a special area to receive and store this part, and a special cart to move the part around in the assembly line. The operator or assemblers have to wear face masks, gloves, and are allowed to use only the assigned tools. Even the screw drivers were had a special cover to prevent touching the part by mistake that would leave a scratch on this housing that may eventually crack the housing. The assembly of the project was in full swing, and things were going perfectly.

My manager called me and said, "Bob, we have a problem with the housing for landing gear control."

He was not specific about the problem, so I asked him more about the problem and he did not answer. I started from the machine shop where they machine the housing, walked through the assembly line and asked the assembly operators, went to the printed circuit board fabrication area, and finally went to the test area and picked a few of the completed control landing gear assembly and operated manually and went back to my manager and reported my findings. I told him that if I were the customer for that control landing gear assembly, I would not buy it because I see a potential failure.

Then he asked me, "Why do you say that?"

I said that is the impression I got when I tested manually few of the final assembly from the test lab.

He said, "Come, let us go to the director of quality."

When we just got into his office, the director said, "Bob, you take care of this control landing gear problem and if anyone points a finger towards to you, let me know, and I will cut that finger off."

I was shocked when I heard that, and I did not get a chance to ask all about the control landing gear. My manager pulled me from the director's office, and he started telling me the true magnitude of the problems.

My manger said, "We had two failures of the control landing gear assembly, one in San Diego, California and the other one in Lemoore, Northern California, while the F/A-18 was taxing in the deck of the ship. One of the landing gears were pulled out and the airplane landed on its wings, but there were no major damages to the airplane because it was in a low speed. The US Navy put all the F/A-18 E/F on AOG and we are in big trouble, since we are the manufacturers of the control landing gear assembly. We have one month to report to the US Navy the corrective action we are taking to prevent such failures."

I was amazed and shocked by this news and they put all the responsibility on my shoulders to take corrective action? I have no knowledge on this product at all and I felt that the management put that (very hot potato) burning issue into my hand. I paused for a while and told my manager to take care of the MMS program and I will investigate the control landing gear issue.

I went back to my office, put all my thoughts together, sketched a plan of action for this burning issue, and asked God to guide me to handle this issue. My next action was to call a meeting. I invited all the department heads such as the director of engineering, the director of manufacturing, the director of procurement, and I was representing the director of quality. We discussed the intensity of problem and they all knew it any way. I gave the assignment to bring all the abnormalities, if any, at the supplier, drawing issues, and any manufacturing issues internally. We met weekly to assess the root cause of the problem. Since there was no historical data of such failures, application of statistical evaluation is out of question, and I was not interested in for any data at all. I was reviewing from the screw to all the components that goes into the assembly of that control landing gear.

There was an office in that company called MRB cage (material review board) where all the rejected hardware is quarantined and waits for the final disposition from the customer. In that office there was one engineer, one customer representative, and another person to handle the materials according to the disposition from the customer. Generally the dispositions were, use as is, repair, or rework – with instructions how to repair or rework the rejected part(s). That is the routine work activities in that office. There will be a big crowd in the morning, all the production control people go there to get their quarantined material and get the non-conformances fixed and continue with the production. Since I was involved in the control landing gear issue, I had the opportunity to go to MRB cage and bail out the rejected parts for the control landing gear assembly. One thing I noticed, after reviewing the nature of all rejections, was that the configuration management was not functioning properly and that contributed about ninety percent of the rejections. I discussed this matter with my director, and I took charge of signing or releasing the engineering drawing for quality engineering.

The routine of configuration control management works like this, the manager at that design engineering department collects all the drawings ready to be released and gets the signatures from engineering, manufacturing, procurement, production control. The final call is for quality to review and sign the drawing for further processing. When I get the stack of drawings, I will review all of them to make sure that completeness of the engineering and most of the drawings were rejected due to noncompliance and ask the design engineer to correct the problems. That process continued for eight to nine months, and I noticed that the activities at the MRB cage reduced to almost zero because there were no more rejections.

My working hours increased to nine to ten hours daily. I go there early in the morning, go through the machine shop, assembly shop, printed circuit fabrication shop, receiving inspection, and finally the final testing area to collect notes and get prepared for the weekly meetings and answer to the DCMA (defense contract management authority) reps. In the morning, a team of DCMA personals will come to the facility and they go all over the shop and start paging me to answer their concerns and doubts and it was a difficult task to answer. I had to run from one end of the company to the other according to where the call came from, and I gave a proper answer to each one of them and they were happy.

Every week Monday morning, a room full of army/navy personnels from the US Navy and Canadian Navy both civilian and naval officers and managers from our company and my job was to update them the progress of the project.

I had full freedom and command to get things done the way I wanted in the right way. There were several milling fixtures, inspection fixtures at the receiving inspection and at the suppliers. They were found defective and removed from use and redesigned newer ones and released with proper instructions on how to use them. Several inspection instructions were created and trained the inspectors how to use them properly. One of the responsibilities I took charge of was to control the manufacturing and assembly planning and procedures. Planning brought to my review was rejected and I walked to the manufacturing engineering area talked to the manager of that department and told him that the planning is not acceptable and asked the manager to send the responsible manufacturing engineer to my office to review the document and make corrections to make it a quality document. When I corrected the manufacturing planning, it was so easy for the people on the floor to follow each operation and there was a tool or gauge to verify the completed operation complies. The operator signs and stamps the document and moves to the next operation. Another taunting issue was one of the purchased components, that was Solenoid, that goes into the assembly and found that was not meeting the requirements to properly function the control landing gear assembly. I discussed this problem with the officers during the Monday's meeting and got the specification revised to make the quality and reliability of the Solenoid fail proof.

Special tools and gauges were designed by me, qualified and created procedures to use those gauges and tools during the receiving process of the Solenoids, same set of gauges and tools were provided to the manufacturer of the Solenoid with proper instructions on how to use them. Several quality policies and procedures were created to make the fabrication of the control landing gear assembly more reliable, and quality built into it to make it a fail and idiot proof product. New products were subjected to all sorts of environmental tests such as a sand and fog test, a vibration test and any other requirements demanded by the customers.Finally, after twelve months of hard work, one fine morning in the presence of all the US Navy and Canadian naval officers including civilian officers and engineers, I handed over a fail proof, fool proof, and idiot proof completed control landing gear assembly to the team and asked them to test it according to the accepted test procedure. They were very happy that the project were completed way ahead of the estimated time and they appreciated the work I did on this project so successfully.

One of the officers asked, "Where did you learn all about the control landing gear?"

And my answer to them was right here and I assured them that if the current configuration remains and follows all my instructions and procedures, this will never fail. But if you change anything from the documents, please consult with me to assure proper functioning of the product.

It was no problem to identify and correct technical or product failure problems, but to fight the politics was the biggest challenge.

When all the critical problems were solved, every single employee in the company was incredibly happy and they were expecting some sort of promotion for me and I was also thinking in that direction. But some strange things happened, my immediate manager was fired soon after I resolved the issues on the control landing gear because he was the in charge for that project for a long time and he could not do anything to prevent such a failure. At that time, our office had about six engineers at different assignments and gradually the total staff was reduced to two. Now our office had only two engineers and one senior manager, and the office was functioning much more efficiently than before. I never had any discussion with my director regarding my promotion; I was waiting for it to come out of his mouth.

Almost year and half passed and tension started building up within me and suddenly it appeared that the receiving inspection supervisor was promoted to the manager's position. That action from my director made me a little upset and one day I had my resignation letter typed and ready for discussion with my director. The following day, I told the secretary to take messages for the director because we are going to have a meeting.

I started my discussion politely, wishing him a good morning and asked him how things are going in the quality department and the overall customer satisfaction on the F/A-18 control landing gear project, and he responded all going well. My next question was regarding the filling of the manager position by promoting the receiving inspection supervisor and I asked him the reason why I was not picked for that position, maybe my race is something that does not qualify me, I am overqualified, or is it just discrimination? The moment I saidthe word "discrimination," he immediately insisted that it was nothing like that. I meant the discrimination of race, color, and my knowledge. So, I changed that discrimination to being against my knowledge. After a long discussion, finally he agreed that he was concerned about my critical thinking and skill in managing projects successfully. He agreed that he was concerned about my knowledge, that by promoting me to the manager's position, sooner or later I may jump into his (director's) position. This discrimination against color, knowledge, and ethnicity will continue if the company policy remains loose and with no justification to protect the resources of the company, knowledge of the employees are the resources of a company. In my case, the HR department did not do anything to investigate the reason why the promotion wasn't given to a person who saved the company from being locked out by the US government. At that point, I explained the whole history. How he told me to take the failed control landing gear project and I never asked for any reward if I do take care of the problem successfully. I thought I would be rewarded appropriately and he has cheated me by offering that job to a white person from the receiving inspection.

I pulled my resignation letter from my pocket and gave him with the remark, "Look I have been faithful to you and to this company, saved the company from being shut down by the government, and the number of staff has reduced more than fifty percent and still the whole QA department I was running efficiently than before. So I cannot trust you by word and by deed. Now I have no intention of working for you and I am leaving. This is my resignation."

He did not say a word and quietly took the letter from me.

Soon after I came out of the office, the information spread like a wild fire throughout the company and the directors and managers from other departments came and requested me to stay and I thanked them. The entire staff and employees were shocked. They felt sorry and some expressed concern over whom should they go with any question or help? I responded that they were their own from now on and that was the end of my stay with Mcdonnell Douglas Astronautics Company.

Chapter Five

Baxter Healthcare Corporation, Edwards Division and Paramax Division.
QA Staff Engineer / Quality and Reliability Engineer.

After working in the aerospace industries, I decided to move to a different industry and the opportunity came to apply for a job with the Baxter Healthcare Corporation, Edwards Division in Irvine, California. I mailed my application and after a week, I got a call from the office to attend an interview. The interview went very well and they told me that they will contact me later with the results. After few days, I got my appointment letter to the position of QA staff engineer.

I reported to work as mentioned in the appointment letter and my manager took me around the office, production floor, receiving inspection department, and shipping area. I was in close contact with my manager to get a status of the quality of the products manufactured and any issues related to the products. My manager explained the various products manufactured in that division and they were: IV pumps, pasma machines and chemistry machines. I went through the floor rejection data, receiving inspection rejection data, and the customer return or customer reported products issues.

After reviewing all the data available at that time, I saw a big communication gap between the design engineers and that reflected all the way through production and testing of the finished products. I reported my findings to the manager with suggestions to improve the situation. My suggestion was to train the engineering design personnel on the fundamentals of engineering design and the GD&T Standard, that standard controls the features and the languages of mechanical engineering design criteria. Upon review of my report, my manager took me to the company president's office and explained the situation of the working conditions based on my report and the requirements of training all the engineers. Upon listening to us and looking at my report, he immediately gave the charge number for training all the engineers, that included design engineers, manufacturing engineers, quality engineers, and the receiving inspection personnel.

We contacted training companies from outside and we got few agencies, and we picked the best one according to their performance history. After the training, we noticed significant improvement in the work performance of the employees and the number of floor rejections went down and the quality of the products was also improved.

One of the critical issues came from hospitals all over the world regarding the massive rejection of thermistor connectors. This is the working process of the thermistor assembly. The thermistor assembly consists of two components, one male unit with a beryllium copper pin and one female unit with a beryllium bushing pressed in the center. When these two are put together, the male unit connects with the female unit and that is connected to the patient monitoring unit.

Problem reported: When the hospitals tried to connect the male and female together, it would not connect or meet each other, this problem was reported from most of the users from all over the world.

Investigation: Because of the urgency of the matter, I was given the project to investigate the problem and take appropriate actions to eradicate the reported problem.

1. My initial investigation was to find out the makers of these male and female connectors. One of the components, the manufacturer of the female part of the connector was in Riverside. I contacted them and made an appointment for a factory visit. According to their management, everything was going well and they were never told of any issues with the part they were making. The shop visit revealed that the part

is made by injection molding process, when the part comes out of the machine, they remove the flash, if any, and they install the eyelet inside.They never checked the alignment of the eyelet, and they were not instructed to do that check either. It is confirmed that they have an issue of checking the alignment of the eyelet.

2. From the procurement department, it is understood that the male part of the connector is made in some part of South America and after making those male connectors, they are shipped to a location in Puerto Rico and this Puerto Rico location receives the female assembly from California and makes a kit of the male and female together and ships to all the hospitals throughout the world. Since the location of the male connector was in South America, I decided to call for few samples from the manufacturer for the evaluation process. Review of the male connector has revealed that the installed pin was not straight. The main problem was the alignment of the two components of the connectors.
3. ASSESSMENT AND CORRECTIVE ACTION. I designed two different gauges for each connector to verify the alignment of the eyelet and the pin and give those gauges to the manufacturers of each connector.

I designed two gauges, each made of steel, one to check the alignment of the eyelet on the female connector and the other to check the alignment of the male pin. I identified those inspection gauges with their part number with the dash number and created the instructions on how to use the inspection gauge during the production of the connectors. I sent those gauges to the suppliers with inspection instructions and asked them to ship the first batch of five parts each to my office for verification of the alignment before I give final clearance to go ahead with the production.

The first batch of the connectors was received from the suppliers and I verified the connectivity of each and found all those five sets of the parts were connecting without any problem and the final clearance was given to each supplier to commence the production and make delivery as soon as possible.

I instructed the Puerto Rico warehouse to do a batch inspection by connecting the male and female units to see the easy fitting of them together and asked them to report if there was any problem to connect them together and to send the bad parts to my office for verification. No complaints were received from any hospitals and the corrective action was extraordinarily successful.

PROBLEM WITH THE CHEMISTRY MACHINE

Things were going smoothly at the Edwards Division for some time and my manager called me to his office and told me a confidentially that they're going to be a big layoff of employees. There would be security guards around to control the employees affected by the layoff and that I was OK, nothing to worry about.

The following day when I got to the office there were security guards to control the employees going inside the company, and I was allowed to go inside. This is what happened there, they let all the R&D engineers go. That is why the security escorted them to their office to collect their belongings and let them go out without touching their working computers.

Background of the Chemistry Machine.

This chemistry was designed at that time the AIDS disease was in full swing and this machine helped the handling of blood samples from the patients during the blood testing process. Nurse draws the blood from the AIDS patients, seals the test tube with proper identification, and sends it to the lab to get tested. The test lab receives this blood sample and places it into the machine at the proper location and turns the machine on. The machine does the rest of the operation such as an automatic robotic arm with a drawing needle punctures the blood sample, raises up, swings back to the cuvettes (it is a clear plastic tiny test chamber) , and drops the blood into each cuvette. The robotic arm then goes back to its original position and the machine does all other operations automatically and this machine is capable of doing around 120 chemistries in a batch. It is very powerful and there are all kinds of technologies in that machine.

During the process of drawing the blood from the test tube, the robotic arm goes down to puncture the test tube seal from the top and inserts the needle, draws the blood, raises the arm, and transitions to turn to the cuvette. It makes a ninety-degree turn before lowering the needle with blood to the cuvette. Blood splashes all over, including the operator, during the ninety-degree turn and that defeats the designed intent of the machine. This was a major issue for the company on the sale of this machine and complaints from the labs those who owned this machine.

This is the biggest issue now. They let all the R&D engineers go and my manager asked if I could help to identify the problem and recommend a fix. I took the responsibility with the help of a manufacturing engineer. With the help of the operation manual and the design manual, I started tracing each assembly and its function. When evaluated the robotic arm and the gear that makes the ninety-degree turn. I found that design tolerances were too wide on the busing internal diameter. That was the reason the blood was splashing when the robotic arm rotated ninety-degrees to reach the cuvette.

Corrective Action: Changed the internal diameter of the busing to a remarkably close tolerance of 0.0002" to 0.0005" and when the new bushing was installed and put the assembly back on the machine, there were no flashing of blood any longer. This corrective action implemented to all the production units and the units in the field.

Things were going smoothly for some time and my manager called me to his office and told me that he is being transferred to Baxter Paramax Systems Division, Irvine, California. Then I told him that the situation in this division is getting little scary and I asked if I would get a chance to move to Paramax Division and he responded to stay back and when he moves to that office, he will take necessary steps to move me from Edwards MDS System to Paramax Systems Division.

When the headhunters keep calling you, that signifies something is going to happen to that division and people keep looking for jobs elsewhere. And that was the time I made my decision to look for a job referred by the headhunter that was Philips Ultrasound, Santa Ana, California.

Chapter Six

Philips Ultrasound, Division of North American Philips Corporation
Santa Ana, California.
Sr. Quality Procurement Engineer.

I joined the Philips Ultrasound Division as a senior quality procurement engineer and my main assignment was to control and manage the supply chain making parts and assemblies to this division. The main product the company made was the ultrasound scanning machine.

This company had many issues with their suppliers, quality of the parts, and assemblies and subassemblies were extremely critical for the successful operation of the scanning machine.

Some of the major issues were:

1. Printed circuit board manufacturers for R&D purpose. The company orders a larger quantity of PC boards with some of them about ten to twelve layers, through hole and surface mount boards. Engineers use only one or two of such boards and scrap them after their R&D work. During their development work, they scrapped hundreds of PC boards. When I noticed the alarming rejection of expensive board, I informed the president of the company and his answer was, "Just ignore it." That is the cost of developing new products, and I never reported such a waste of money.
2. Supplier made PC boards were not meeting the requirements of

specification. I launched an investigation on the suppliers who are responsible for the fabrication of the PC boards and found that it was not the supplier's fault, but the components for the fabrication of the boards, the kits supplied by the Philips Ultrasound were bad, not to the pick list according to the engineering drawing. I coordinated with the manufacturing engineer responsible for kitting the electronic components and made changes in the process of picking the right components for each kit and made accountability for those who worked on that process.

3. Major issues with the hybrid assembly supplied by the manufacturer in South Korea, eighty to eighty-five percent of the hybrids were rejected during the testing of the ultrasound machines. This alarming rate of rejection was a major concern of the company and I launched an investigation on this issue.

I contacted the procurement department to find out the supplier's address and contact information in S. Korea and the local contact point in California. I also investigated the status of the overall working conditions in S. Korea and learned that the country/company had a major strike and most of the manufacturing companies were under lockdown for more than six months. I was curious to know more about the contractor in California. With the help of the procurement department, I got the contractor's name and phone number and made an appointment to visit their shop here in California.

The buyer from procurement and I visited that company in the Irvine area. The owner welcomed us into his office. We had a brief discussion regarding the failure of the hybrids made by the S. Korean suppliers and mentioned the employee strike in S. Korea. We asked him how he controls the products from that supplier. He mentioned that the company was shutdown for more than six months and they shipped all the components with partially finished hybrids to the Irvine office and he took us to the garage where all the parts and subassemblies were dumped. He makes the hybrid assembly by himself by soldering components and ships the hybrid assembly to Philips Ultrasound for use on the machine. We thanked the company owner for showing the operation in his garage, and upon arriving at our office, I called for an immediate meeting with the finance, procurement, manufacturing, and quality to address the situation. At the end of the meeting, an emergency fund was created to travel across the United States of America to find a manufacturing facility capable of making the hybrid assemblies. Purchasing has collected several supplier's contact information and our journey started in Baltimore. We visited a few suppliers there and travelled to other states and finally we landed in Wisconsin at another division of Philips Corporation. They make all kinds of PC board fabrications and we reviewed their capabilities and the quality performance. Our team has decided to approve this Philips division to make the hybrids for our need. The first batch of the hybrid assembly was found acceptable and the performance of those hybrids were found superior to other hybrids made by Korean source.

KEY POINTS TO REMEMBER WHEN OUTSOURCING, ESPECIALLY OUTSIDE USA.

It is particularly important to understand the suppliers you are selecting to do business with in other countries. I hope the following points will help you in supplier selection.

1. Understand the language barrier and if they understand English fluently, make communication effective.
2. Verify their manufacturing capabilities for the products you are trying to negotiate.
3. Do not send a statistician or person with Six Sigma, SPC guru for the evaluation of a manufacturing company. Rather, send someone from the manufacturing department along with a quality engineer who knows the product and processes.
4. Evaluate the company has a good machine maintenance program to keep the original machine manufacturers specifications to the specified limit.
5. Check their quality performance with their existing customers.
6. Understand the political situation of that country by checking the history of the employee strikes and company lockdowns in the last five years. If the company had labor problems in the past, avoid making a contract with them.
7. Verify the way they communicate internally with the workers, assemblers, and machine operators with proper documentation to validate the accountability of the operators for the work they do.

8. Make sure that the company has a good quality management system. ISO and third-party certifications have proven not that effective as far as the product quality is concerned. If the company has a good quality culture, that would be much better than all those expensive banner type of certifications. If they have second party certification it is much better than third party certification and ISO and similar certifications.
9. Look at their MRB system and the reports for few years or the management minutes of meetings, from that you may be able to determine the performance of the company.
10. Study their business ethics and see how long they were in the business, that will tell the stability of the company and their quality performance.

Do all the above-mentioned things before you sign any contract with the new suppliers in any part of the world. Do not conduct business with manufacturers in China or any Communist countries.

Chapter Seven

Clayton Industries, El Monte, California
Quality Assurance Manager.

I had no desire to move away from the Philips Ultrasound in Santa Ana. In our management meeting, the president of the company, brought up the loss of several millions of dollars and my answer was to investigate the trash bin where they have scrapped hundreds of expensive PC boards with ten to twelve layers of boards made for R&D projects. One fine morning, corporate has decided to lay off the president of the company and from there down big officers. I figured out that my turn would be next and when the opportunity came, I moved to Clayton Industries with the help of employment agencies.

I was very new to the products made by this company; they make steam generators and I never knew the wide application of steam generators. The manufacturing process of the steam generators is very strictly controlled by the ASME inspectors because the critical application of the product where quality and safety were concerned was prime.

When I went through the manufacturing floor to understand the culture of the people and the processes, my findings were very shocking. They used to make simple plain steel washers and many hardware items in their shop Their manufacturing culture was centuries old, while across the street, companies had the latest technology. I cannot change their culture within a week or so, but I brought my observation to the upper management and to the president of the company.

The product manufactured was the steam generators. There was a government inspector that monitored the production of the steam generators and the final buy off is done by the inspector. Without their acceptance stamp, the company may not be able to sell the product.

I made two significant contributions to that company. Number one was to streamline the assembly and testing of manufactured products and made more transparency and accountability of the assemblers and the inspectors on the job. The government inspector appreciated the work I did, making the whole process very transparent and accountable.

My next action was to reduce the cost of operation and control the process at the Mexico plant. The manufacturing process of the coiling was done at the Mexico plant. They received ERW steel tubes from the manufacturer and pick from the hanger and move to Mexico and dump the tubes there in the sand. It may lay in the sand for a maximum of six months, then the tubes are taken to caustic etch process and dipped in the tank until the grease is totally removed from the tube and they go to the coiling process. After coiling, those are moved back to the El Monte, California factory where the rest of the operations are completed until it is tested and delivered to the customer.

PROJECT IMPROVEMENT AND QUALTY

I was interested to increase the quality of the product and reduce the cost of operation of the coiling process, right from the raw material to the coiling of the tubes in Mexico.

We order the ERW tubes from the local supplier in Southern California and they oil it to protect corrosion and keep them in their hanger until it is drawn and shipped to Mexico plant. When we need them for coiling, our truck pulls them and transports them to Mexico and dumps them in the sandy land and they stay there for six months or so in the open weather, rain and heat, until the tubes are pulled for cleaning and the coiling process and then send them back to the California plant for further processing. During production, they pull the tubes from there and dip them in the caustic tank to degrease them. Once it is degreased, that goes through the cleaning process and to the coiling process.

My effort was to reduce the cost of operation and improvement in quality of the product, and I found the opportunity to do that.

First, I contacted the ERW tube manufacturer and asked them, "If I ask you to eliminate the greasing the tubes, will there be a cost reduction?"

They said, "Sure," and I asked the procurement to have the cost reduced to whatever the manufacturer says.

The next step was to collect a rusted cut piece of tube from the sandy area in Mexico. It might have been there for a long time, exact duration is not known, but estimated, it should be around six months old at least. I collected another cut piece of the tube from the new batch and identified those two pieces of tubes and sent them to a testing lab to check the samples for their degradation with the rusted one against the new one. The test results came out that they could not find any difference between the rusted one and the new sample.

I discussed this matter with the upper management and the government customer and informed them that I was going to skip greasing or oiling the tubes immediately after manufacturing them. The data shows that the tubes would not stay in the hanger for more than six months and the purchase order was amended to skip oiling the tubes. The next step was to eliminate the caustic etching of the tubes. That saved lots of time and the tubes went straight to the coiling process. Later, the caustic tank was eliminated. That was a big relief for the company in Mexico because the Mexican government was not happy with the use of caustic tank that would damage their environmental condition and they were happy the caustic etch tank was removed.

The total cost savings was close to a million dollars per year. All these things happened within six months with that company, then I got sick and was admitted to the hospital for an appendectomy. Company management was cheap and they laid me off. I had to stay in the hospital longer due to an infection.

Chapter Eight

Quality Engineering Management Consultant.
Self-employed.

After leaving Clayton Industries, I started a quality management consulting business. I had two clients, one was in Oxnard, California. The job assignment was to create a complete quality management system and get them certified to ISO 9001. The product manufactured by the company, Lumonics Corporation, was a laser engraving machine. I created a complete quality manual and separate work instructions for the control of the manufacturing process of their laser engraving machine, an employee training program, and receiving inspection procedures.

The second customer was the automotive component manufacturer Trim–Lok, Inc. in Buena Park, California. They make extrusions for the trims for automotive vehicle applications. They design and make different varieties of trims in their own shop in Buena Park, California. They never had any certification or any quality manual or procedures to control the process. One of their customers was the Volvo Corporation and they wanted Trim-Lok, Inc. to get certified to ISO 9001. At this point, the president of Trim-Lok called me to help them to get this situation discussed and do the necessary procedures to get their approval to do business.

I had a meeting at their office with the president of the company and the director of production. After discussion of the certification process and the cost involved, the president of the company suggested coming up with an alternative solution for the ISO certification process. It may be noted that this company never had a quality manual or any procedures for outsiders or customers to see and assess their strength in quality or manufacturing process or both. At that point, I asked their permission to do an assessment of the process control and the system of how they control the quality of their products. They had one process control inspector and he reported to the director of manufacturing and that was it.

After my internal audit and assessment, I presented my findings to the president of the company and he suggested me to be present in their meeting with the representatives from Volvo Corporation of Sweden as a consultant and listen their requirements and do the required policies and procedures at a later date.

When the representatives had their presentations and their expectations from Trim-Lok completed, they gave their findings along with a corrective action to be completed within a time frame. I had meeting with the president of the company to discuss the future and how to proceed with creating the quality manual and work instructions etc. Upon listening to my suggestions to go for a third-party certification, hiring a quality manager, and creating a quality department, the president of the company suggested that I to come up with an alternative plan to reduce the cost of the entire project.

My suggestion was this, (1) Create a corporate quality manual specifying how the company is involved in controlling the manufactured products and how do they control the process of making the trims. (2) Create work instructions to explain the details of the manufacturing process for each product, including shipping the final products to their customers. That suggestion was approved by the president of the company, and I started on that project immediately.

I created the company structure like this: president, vice president, director of engineering, director of manufacturing, director of procurement and director of sales, quality process auditor and inspector. Note: you may see that there is no quality department, but they have a quality process auditor and inspector which is one and the same. What I did was to assign each department with the responsibility of quality function, and they are responsible for those activities and the quality inspector does quarterly audit in each department and reports directly to the president of the company. The president holds each department director responsible for the correction and the inspector follows up the implementation of all the corrective actions and reports back to the president of the company with the results.

When I presented this company structure, he liked it and assigned me the task to implement this system throughout the company. Accordingly, I created their first quality manual and work instructions for each product and inspection criteria to guide the inspector to do his daily activities and created audit checklist for each department and a reporting document to the president of the company.

This company used to get lots of quality survey questionnaires from various would-be customers/companies to assess the capability of this company. The procedure I created was to send them the companies' quality manual and if they are not happy with the quality manual, then charge for each page, the first page was $200 and subsequent pages were $100 each. Most of the companies were happy with the quality manual and any additional questions were answered by email or letter. I was their consultant, working from offsite, answering all their technical and quality questions and inquiries through mail or email and attending their monthly meetings at the company's office. I continued this task for few years until I was selected to go back to the Boeing Company in St. Louis Missouri. Trim-Lok wanted me to continue as their consultant and I refused to continue, that would be a conflict of interest to the Boeing Company and especially when I would be dealing with defense contracts and suppliers.

Chapter Nine

BOEING COMPANY, INTEGRATED DEFENSE SYSTEM. St. Louis, MO
Principal Specialist Quality Engineer, Supplier Quality, CA Region

[CAUTION: From my experience with the Boeing Company, both in Mcdonnell Douglas Astronautics Company, Monrovia, California, and Boeing IDS St. Louis, MO. The quality department is the most corrupt and discriminatory with injustice to minorities, treating them like slaves to get excellent work done by them and not giving any financial compensation or an annual pay increase for work well done. They do not even have a pay increase to match the living cost index. It is amazing to say that the human resources department have nothing to say about this heinous, unethical, and horrible crime of the quality management does to their employees and they just follow the blueprint agrees with quality department says.

I have seen in other companies the HR department is the strongest one to protect the resources of the company, that is the employees, they play very important role in keeping the qualified employees with the company and reviews their work performances closely to make sure their immediate management treats them equally and give salary increase on merit according to their performance.

It is evident that the Boeing Company have a mean and corrupted management system that should be changed. A company of this magnitude and reputation should change the management style of managing the employees and should treat them fairly and give promotion to the next higher level when the situation arises. HR department should also change their style of treating the employees, they should have their own standard to measure and retain the good employees it is good for the company and the employees would contribute to the company technologically for the growth of the company. Good and productive employees are the resources of the company and HR should have a system to measure, evaluate each employee's contribution to the company and should never take the words from the management the employee directly reports to.]

Principal Specialist Quality Engineer, Supplier Quality, CA Region

While I was doing my own consulting services to the manufacturing industries, I noticed an advertisement from the Boeing Company St. Louis, MO, looking for a self-managed quality professional to manage the supply chain supporting their defense programs and I tossed an application showing my interest for that position. After a few weeks, I received an interview call from Boeing and the location was in a hotel by the Los Angeles airport. As per the schedule, I was present at the hotel and the team welcomed me and explained the process of selection, that is a written test and a panel interview. After completion of the written test, I scored 100% correct, I had the panel interview. After detailed questioning by the panel, they informed me that the office would contact me for further action. After two weeks, I believe, I received the appointment letter from the Boeing Company, St. Louis, MO. And my job title was "principal specialist quality engineer, supplier quality, California region".

This is the way the job assignments work, my office is in St. Louis, MO and my actual working location would be in Los Angeles, California. Job assignments would be directed from the St. Louis office and I would follow their direction and complete the job assignments by completing the executive summary of supplier visits to make the supplier manufacturing process delivering quality products.

It is important to know that the function of the quality department is/was an after-the-fact process, always with reactive quality checks at the supplier. I was wondering how come quality does not act on a proactive quality initiative in supplier management, that was/is their, Boeing's, quality culture. No wonder the products are failing in various functions. When I discussed this matter with the management, they did not welcome my suggestions and I realized that I was preaching to the wrong crowd. That reflected badly on my professional carrier. They used me as a holy cow to resolve supplier related issues efficiently and economically.

There was an employee training program for two weeks in St. Louis that includes taking pictures for the employee badge and completing human resources documentation, etc. Company policies and procedures, suppliers' assignments and timely audit at their facility, control and manage the supplier ratings, identify the need for supplier training or corrective actions to improve their acceptance ratings and report all these to the St. Louis office online.

I believe there were five to six field engineers in Southern California taking care of the suppliers supporting the (defense) contracts from St. Louis, mostly on F/A-18, F/A-15, JDAM, space shuttle, etc.

Supplier management is defined and controlled by a process called PVA, that is process validation assessment. PVA is an after-the-fact process, checking all the documentation, mainly the manufacturing process, and some data from the inspection sheets at the supplier and at the end create an executive summary based on the findings or evidence. Sometimes the supplier might have exercised good process control or in some cases, you may find process deficiency and suggest improvement to the process. In that case, a corrective action document is created and a response is asked for within a specified time. When the corrective actions are verified to see the accuracy of the action plan, that case would be closed.

The irony of the case has not finished there. Management at the St. Louis office go through the PVA document line by line and interpret the language, the way the summary was written, and they will contact the employee and ask for clarification, ask why such an action was taken, and all kinds of interrogations that have no technical merits, but is political in nature. This reflects in the employee's record and at the end of the year, all such findings on the employee will be taken against the employee and they will stop giving them any annual compensation or salary raise.

Management has no technical knowledge of all kinds of processes going on at the suppliers and they purely depend on the PVA summary, the dirtiest and rudest way to manage the employees. At the same time, management shows favoritism towards their employees based on their color, race, and nationality.

From my experience, what I noticed that the so-called quality organization is a totally polluted organization with favoritism, discrimination, racism, and politics and I have to say that their technical knowledge is very minimum or nil. But they have a strong passion on imposing the use of statistical tools and its applications in governing and managing the suppliers for their defense. I explained to the management that it would be better to have a proactive quality management system implemented that would replace the existing after-the-fact quality check and management system. Probably the management did not welcome my honest suggestion and that fired back on me, and I became a noted candidate, subject to scrutiny.

I am a victim of such discrimination and racism as a minority because my knowledge in the manufacturing process at the suppliers and my ability to solve technical issues has never been recognized or awarded or at a minimum, the cost of living has not been given during the annual salary review and forget about chance of promotion. Overall, my experience in the quality department can be summarized as a technical labor camp where you take care of all the technical or supplier related issues and problems and do not expect any monetary benefits or at minimum a salary raise.

Regardless of the treatment of my management, I never showed any discrimination in my professional decision-making process or helping the suppliers resolving the problems. I never consulted with the management on any technical matters of the problems the suppliers are facing because I knew they many not able to give any technical support or assistance.

I would like to bring couple of issues the suppliers were facing and how I got resolved the problems and the suppliers were able to make and ship quality products to the Boeing assembly plant in St. Louis, MO and Long Beach, CA.

1. A problem reported from the Boeing St. Louis plant was that the spar for the F/A-18 aircraft was ¼" shorter in overall length and they could not assemble the plane and that report popped up in my computer and I started to investigate the problems right away. I arranged a meeting with the supplier's management, reviewed the manufacturing plan, material handling, outside manufacturing, and their manufacturing plan, tools, machines, and techniques. I visited the heat treatment plant to ensure proper heat treatment of the parts and heat distribution, validation, and calibration of the oven and thermal cy-

cling, etc. and all those operations were found to be acceptable.

The next investigation was the material (titanium sheet furnished by the Boeing Company St. Louis, MO). The supplier mentioned the material has furnished by the Boeing Company with proper material certification. I reviewed the material certification and found it to be meeting the material specifications. I asked the supplier if they have any raw material from St. Louis, MO available in stock and found that they had six pieces in stock. I asked them to cut two pieces from the raw material and send it for material analysis and physical analysis such as a tensile and elongation test and the test results found that the tensile test failed. That proved that the Boeing furnished material was not in compliance with the requirements and should be heat treated and annealed to the condition specified in the drawing. That means the material certification was fake and I reported this matter to the St. Louis, MO quality department. Their employees did not verify the condition of the material before shipping to the supplier in California.

When a rejection of this magnitude occurs, that goes to the supplier's quality rating and their rating went down to red from green. I contested the quality rating of the supplier and requested to reverse the quality rating to green.

St. Louis, MO procurement gave additional cost to the supplier to do the heat treatment of the material to the condition mentioned in the engineering drawing and the supplier made first two spars, with all the documentations, and sent it to St. Louis, MO for assembly on the plane and found those two spars were good, no more gap when assembled. I asked the supplier to continue the production of the remainder of the spars according to the approved manufacturing plan, with inspection documentation prior to shipping to Boeing St. Louis, MO. and that problem has put to death.

2. Problem reported that the C-17 wing, when assembled, found ripples at the edge of the wing along the edge of the wing. This caused the stoppage of production and action was required immediately to identify and correct the problem.

   I contacted the supplier and made an appointment with the management to investigate the problem and had a discussion with the management. I requested a plant tour to see the manufacturing process of the wing. I reviewed the manufacturing plan, inspection process, and the inspection fixtures provided by the Boeing Company, Long Beach California and I asked for any instructions that came along with the inspection fixture and they said no instructions were furnished.

Deficiency to check the distance between the ribs of the fixture was not addressed or furnished a gauge to check the condition and I asked the supplier to make an inspection gauge to check the distance between the ribs of the inspection fixture, identify part number, the revision of the drawing, the project name, and the date of installation in service and the calibration date. I asked them to make their own inspection instructions and I reviewed and approved the plan.

After implementation of the revised planning change with added inspections, the parts were found to be following the engineering drawing and there were no more ripple effects on the wing surface. The first batch of two wings were shipped to the Long Beach facility and they were happy to see the problem go away. Upon their satisfaction with the product, the supplier continued the production of the wings with 100% inspection data furnished for my review, and the problem never surfaced again.

I was also responsible to manage and control the supplier making critical coaxial cable assemblies for the space lab/space shuttle applications. When I took charge of that supplier, the previous person managing that company's process management had a very long process that takes five days a week or longer depending on the product. Product quantity varies, mostly one part per contract line item, and like that several products need to be verified to see that they meet the requirements.

When I took charge of the product management, I trained the supplier with instructions on how to build quality into the process, document the process, and document inspection data for evaluation. I made the supplier's quality manager responsible for the inspection of all the cables and to sign the document before submitting for source inspection. The supplier was very happy to follow up with the system I made, and they were able to ship 100% quality products on time and the source inspection time has reduced about fifty to sixty percent.

During management discussions, one of the engineers managing the exact same product at a different supplier had problems evaluating and clearing the parts for shipment. I explained the reasons to the management why there were delays in processing the parts:

1. Systemic problems, source engineer follows the manufacturers system of inspecting the parts over and over, that causes delay.
2. Part processing or accepting process is way higher than the normal situation and the inspection cost goes very high.
3. Source engineer does not have the ability to manage and define the controls of the high tech coaxial cables for defense applications.

It is very clear that the management also did not get the points I was talking about, and they did nothing.

I see the irony of the situation, one company with different quality standards, because of the lack of knowledge, management decided to close their eyes and relax in their office. This is one of the examples why the Boeing company's quality performances went way down, and the product failures are the results of ignorant managements actions at the top level.

The End

QEM Consultant

Logo

INDUSTRIAL STATISTICS. BIRTH OF A NEW TECHNOLOGY A HYBRID TECHNOLOGY.

Birth of Reverse SPC Software.
Key Technical working elements of the Industrial Statistics.
Origination of the Reverse SPC.

Before we talk about the Reverse SPC (reverse statistical process control), we must understand the basic working methods of traditional statistical process control, where it started, and how that method spread across the globe and how that affected small and big industries across the globe.

Let us examine where the statistical process control started from. First, search Google, ASQ, and Wikipedia and you may see the results below. The history of statistical process control is only to give you an idea to understand the What, When, Where, Why, Who, that is all, nothing beyond that.

When: 1920 – 1924
Where: Bell Laboratories
Who: Walter Shewhart
Why: Process control (Control chart)
What: State of statistical control

Google search: Statistical process control was pioneered by Walter A. Shewhart at Bell Laboratories in the early 1920s. Shewhart developed the control chart in 1924 and the concept of a state of statistical control.

ASQ search: Statistical process control (SPC) is defined as the use of statistical techniques to control a process or production method. SPC tools and procedures can help you monitor process behavior, discover issues in internal systems, and find solutions for production issues. Statistical process control is often used interchangeably with statistical quality control (SQC). Statistical quality control (SQC) is defined as the application of the 14 statistical and analytical tools (7-QC and 7-SUPP) to monitor process outputs (dependent variables). Statistical process control (SPC) is the application of the same fourteen tools to control process inputs (independent variables). Although both terms are often used interchangeably, SQC includes acceptance sampling where SPC does not.

Wikipedia: Statistical process control (SPC) is a method of quality control which employs statistical methods to monitor and control a process. This helps to ensure that the process operates efficiently, producing more specification-conforming products with less waste (rework or scrap). SPC can be applied to any process where the "conforming product" (product meeting specifications) output can be measured. Key tools used in SPC include run charts, control charts, a focus on continuous improvement, and the design of experiments. An example of a process where SPC is applied is manufacturing lines.

SPC must be practiced in two phases: The first phase is the initial establishment of the process, and the second phase is the regular production use of the process. In the second phase, a decision of the period to be examined must be made, depending upon the change in 5M&E conditions (Man, Machine, Material, Method, Movement, Environment) and wear rate of parts used in the manufacturing process (machine parts, jigs, and fixtures).

An advantage of SPC over other methods of quality control, such as "Inspection", is that it emphasizes early detection and prevention of problems, rather than the correction of problems after they have occurred.

In addition to reducing waste, SPC can lead to a reduction in the time required to produce the product. SPC makes it less likely the finished product will need to be reworked or scrapped.

There are many questions you may have when talking about the statistical process control and some of them are given below, extracted from the web site

Who developed statistical process controls?

Walter Shewhart

The year 1924—at a factory in Cicero, Illinois—saw the start of two of the most important developments ever in managerial thinking. In May that year, Walter Shewhart described the first control chart which launched statistical process control and quality improvement.

When was SPC first used?

SPC Tools. A popular SPC tool is the control chart, originally developed by Walter Shewhart in the early 1920s.

What is the idea of statistical process control?

Statistical Process Control (SPC) is an industry-standard methodology for measuring and controlling quality during the manufacturing process. Quality data in the form of product or process measurements are obtained in real-time during manufacturing.

Why is SPC needed?

The SPC process is implemented to move a company from detection based to prevention based quality controls. By monitoring the performance of a process in real time the operator can detect trends or changes in the process before they result in non-conforming product and scrap.

When should SPC be used?

When a process needs to be controlled and 100% inspection is not desired or feasible, then SPC is reasonable. Statistical process control costs are considered failure prevention activities. It cost less than internal and external failure costs.

How does SPC improve performance?

SPC techniques are useful for measuring the value of a quality characteristic and helping identify change or variation in the quality of either the product or process. SPC doesn't eliminate variation; instead, it makes it possible to track special cause variation.

Review one of the examples of X-bar and R-chart copied from the website to show the type of SPC chart used in the manufacturing industry, this is for educational purpose only.

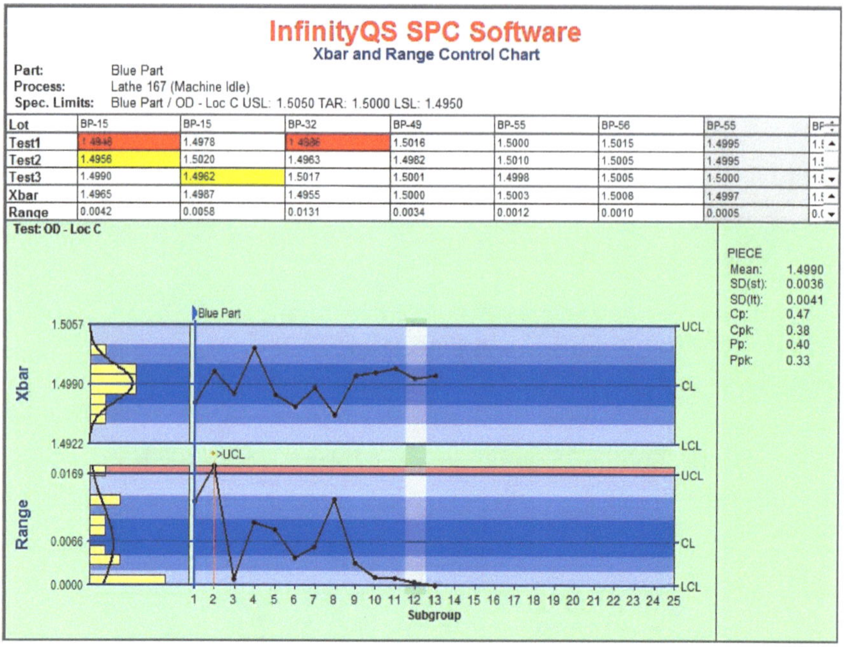

X-bar R-chart taken from the web site for educational purpose.

The idea of creating a control chart like this is to collect data (measurement data from a manufactured product) and plot a chart to give a visible presentation of the behavior of the data or to measure the variations of data over a period.

From this chart you may see that the Name of the Part – Blue Part

Dimension: OD, USL=1.5050" & LSL=1.4950" – This detail is from the engineering drawing.

Target dimension chosen to setup manufacturing = 1.5000"

Machine used is lathe #167

From the chart you can see that a few dimensions are marked red and some yellow.

BP-15 through BP-55 are the lot numbers of the blue part made at different intervals.

Test 1, Test 2 and Test 3 are the tests performed at certain intervals.

X-bar is the average of the dimensions measured and documented.

Range is the average of the Range of dimensions documented.

The right hand side of the chart shows: UCL-Upper control limit, LCL-Lower control limit and CL-appears the mean of the control limits.

And the extreme right-hand side under "PIECE" are the mean and standard deviations of the dimensions from the table and the rest, you may research from the website to know more about it. But we will be discussing more about mean and standard deviation in the following pages. Attached is the frequency distribution chart or histogram means, how many times a dimension has been repeated from the inspection data. You may create one using the Excel program.

In the industrial statistics we will be discussing only three charts mainly,

1. X-bar chart, 2. R-bar chart and 3. Histogram or frequency chart.

The frequency chart or histogram is copied from the web for education purpose only.

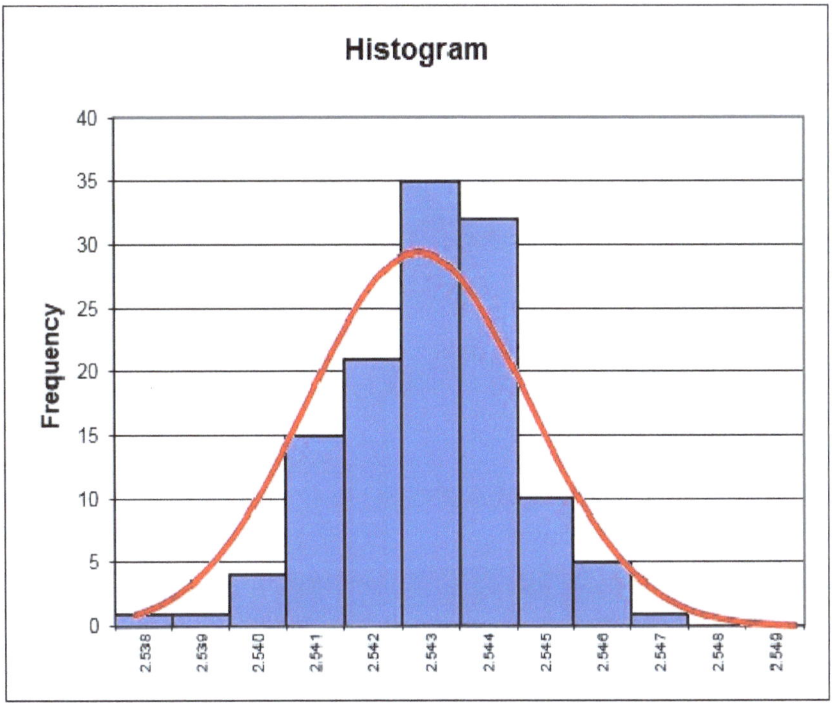

Benefits of SPC

Copied from the web site EQMS for education purpose only "https://www.e-qms.co.uk/"

EQMS

Extended Quality Management Solutions

EQMS Limited

UK based ISO consulting company providing outsourced internal audits, supplier audits and consultant support developing and implementing ISO management systems.

Our consultants are experienced across a wide range of industries, and we support ISO 9001, ISO 14001, ISO 27001, ISO 45001, and ISO 17025.

Statistical Process Control (SPC) in Manufacturing

Statistical Process Control (SPC) is an industry-standard procedure for measuring and monitoring quality during the manufacturing process. Quality data as Product or Process estimations are acquired in real-time during manufacturing. This data is then plotted on a graph with predetermined control limits. Control limits are determined by the capability of the process, though the client's needs determine specification limits. This article will review how to use SPC in manufacturing and some of the benefits it brings to your business.

How to Use SPC in manufacturing

Before performing SPC or any new quality system, the manufacturing process ought to be evaluated to determine the main regions of waste. A few cases of manufacturing process waste are reworked, scrap and excessive inspection time. It would be most useful to apply the SPC tools to these areas first. During SPC, not all dimensions are checked because of the cost, time and population delays that would incur. Preceding SPC implementation the core or essential characteristics of the design or process ought to be identified by a Cross Functional Team (CFT) during a print survey or Design Failure Mode and Effects Analysis (DFMEA) exercise. Data would then be gathered and checked on these key or essential characteristics.

Collecting and Recording Data

SPC data is collected in the form of measurement of an items dimension or process instrumentation readings. The data is then recorded and tracked on different sorts of control charts, based on the kind of data being collected. It is important that the right type of chart is utilized to gain value and get helpful information. The data can be in the form of continuous variable data or attribute data. The data can likewise be collected and recorded as individual values or an average of a group of readings. Some general guidelines and cases are listed below. This list is not all comprehensive and provided just as a reference.

Variable data

- Individual – Moving Range chart: to be utilized if your data is individual values.
- X bar – R chart: To be utilized on the off chance that you are recording data in subgroups of 8 or less.
- X bar – S chart: to be utilized if your sub-group size is greater than 8.

Attribute data

- P graph: For recording the number of deficient parts in a group of parts.
- U graph: For recording the number of defects in each part.

Control Charts

A standout amongst the most widely used control diagrams for variable data is the X-bar and R chart. X-bar represented the average or "mean" value of the variable x. The X-bar charts illustrate the variation in the sample means or average. The Range chart demonstrates the variation within the subgroup. The range is the distinction between the highest and lowest value. The following steps are required to build an X-bar and R chart:

Once the chart is setup, the operator or technician will measure various samples, add the values together then compute the average. This value is then recorded on a control chart or X-bar chart. The range of the subgroups is additionally recorded. The sample measurement ought to be taken and recorded in regular interims, including date and time to track the stability of the process. Look for any special or assignable causes and change the process as important to maintain a stable and in control process.

The X-bar and R chart is just one case of the distinctive control charts available for process monitoring and improvement.

Analyzing the Data

The data points recorded on a control chart should fall between the control limits as far as possible, given that only common causes and no special causes have been identified. Common causes will drop between the control limit while special causes are generally outside of the control limit. For a process to be considered in statistical control there ought to be no special causes in any of the charts. A process in control will have no exceptional causes distinguished in it, and the data should fall between the control limits. A few cases of common cause variation are as per the following:

- Variation in material properties within specification
- Seasonal changes in ambient temperature or humidity
- Normal machine or tooling wear
- Variability in operator controlled settings
- Normal measurement variation

Exceptional causes generally fall outside of control limit or show a radical change or shift in the process. A few cases of special cause variety are underneath:

- Failed controllers
- Improper gear adjustments

- A change in the measurement system
- A process shifts
- Machine breakdown
- Raw material properties out of design specifications
- Broken tool, punch, bit, and so forth.
- Inexperienced operator not familiar with process

While monitoring a process through SPC charts, the inspector will approve that all data points are inside control limits and look for trends or sudden changes in the process. If any special causes for variation are identified, appropriate action ought to be taken to determine the cause and actualize corrective actions to restore the process to a state of statistical control.

Benefits of SPC in Manufacturing

- Lower manufacturing costs
- Improved forecasting
- Decreased issue to solution lead time
- Correct standards
- Less inspection
- Better client relations
- Reliable measures of capability
- Stable processes
- Improved product quality
- Realistic Specification
- Decreased cycle time

Overall then there are many benefits to using SPC in manufacturing which is why it is widely used within the industry. SPC can also be used in many other industries to monitor if a process is in control, for example to determine if the response time in a call center is within statistical control.

The above materials are furnished only to understand the nature of statistical process control, where it is mostly used, and why. When you go deep into this subject, that will take you to a different world and you will be hooked on to that kind of brainwashing process or misguided process, where they will take you to tame you to think in their line of thought. I do not believe in the glory of the SPC program, because it will not work in the manufacturing engineering area and many things, they promise are not true. Real application of SPC will be in an area or process where proper definition of the base, origin, datum is unknown. For example: rain, earthquake, election results, gambling, sickness like Covid-19 and the like.

Manathazathu Principle

"Matthew [Manthazathu] principle says that - in any manufacturing process; if the product is defined, process is known and the variables are known, then use Industrial statistics with R SPC software to plan and control the manufacturing process to yield close to 100% or more, do not use statistical process control, that has a different mindset and focus."

You may have many questions on the applications of SPC, the benefits and drawbacks of SPC; you can search in the website and you will see an ocean of them to confuse you and you may wonder where to go from here. There is no intention to go through all the details of the questions given below, that is totally out of my focus. From now on, most of the discussions will be based on the X-bar, R-chart, and frequency distribution chart.

This process control techniques (SPC) were used in the early days of industrial revolution around after World War II, it was developed by William A. Schwhart and used at the Bell Laboratories around 1924. It was not giving a satisfactory result to identify and correct the root cause of the problem related to products and manufacturing processes. Disadvantage with the statistical process control, it is an after-the-fact process, meaning it collects data from the manufactured products by inspection or inspection data from the inspection data sheet to decide whether the product meets the engineering drawing requirements or is good to consumers or rework to mee the requirements or to scarp and remake them. This process is a heavy financial burden to the manufacturers and substantial delay to get the product into the market. This is an age-old concept when the technology was in the infant stage. Now we are better off, technology is leading the business.

I highly recommend removing the dependance to traditional statistics and installing industrial statistics. It is a hybrid technology, walking you through the process and helps you to tailor quality into the process and the results will be shocking to you. Reduced cost of operation, get products to the market sooner, no more scrap and rework, it eliminates unworthy processes and very transparent. After attending several hours of training in SPC, obtaining Six Sigma Green Belt, that did not make me comfortable to define and control a manufacturing process without any old data. Our late quality guru Dr. Edward Deming said that – yesterday's data will not predict the future – it is very true as far as the manufacturing area is concerned.

Collected data from various manufacturing processes, applied the principles of Statistical Process Control, created X-bar and R-chart and frequency distribution chart or Histogram to analyze the Six Sigma limit and stuff like that. SPC application comes from the sample data collected after the parts are already made. My concern was how to make the manufacturing process robust, so that the product would be a 100% quality product and how can a plan, a road map, or process map help the manufacturer guide through the process of NC or CNC programming. When the programming is done correctly, the operator can check the part frequently and load the data into the R SPC program, that will show the X-bar, R-chart, and Histogram. It is a real-time, visible chart available to the management and to the customers as well.

I took a deep analysis of the data collected for SPC analysis and the concept of famous traditional statistical bell curve and created a mathematical model of a bell curve. It may look exactly like the traditional statistical bell curve, but the properties are entirely different. In traditional statistics, a frequency distribution chart is created and the bell curve is placed on the top of it to see where the data lies, and the determinations are made by the statisticians. It is a cookie cutter process, statisticians play their game plan as they wish. It is a political game plan.

The mathematically created bell curve has the advantage of defining the process parameters or variables in such a way that all the variables have a distinct purpose during the machining process. A derived sigma is also created that signifies the process capability of the machine on which the part is being planned to fabricate. In short, the RSPC software program generates all the parameters required to make a part that meets the customer requirements. For comparison, sigma level before and after the process is available in the program. Frequency distribution charts and X-bar, R-charts are also created from the inspection data input by the machinist, this eliminates multiple inspections of the same part by different inspectors at various points of the journey of the part. In short, it is a full package to define, manage, and control quality of the product from the beginning till the end of the journey of the part or assembly.

"Decades, the manufacturing industries were misguided by teaching the use of statistical tools such as, SPC, Six Sigma, Black Blet, Cp and CpK etc. a technique does not work anymore in an advanced technological world. Instead, this book tells you what and how to plan and control manufacturing process to tailor quality into the process, so that the process will produce quality products consistently". A tool and techniques for the manufacturing industries tomorrow, to manage and control cost and produce quality products.

Several hours of training in statistical process control and statistical quality management have not taught anything technical that guides to control the manufacturing process management and control before the start of the manufacturing process. All the statistical tools taught was after the fact, data from the inspection of the parts already made, this is an age-old technique when the manufacturing process was in the primitive stage; that time has changed, now we are in an advanced manufacturing technological era.

There are several universities and colleges and many consulting businesses, still teaching these dead and nonproductive tools and techniques to millions across the globe. This is wrong, teaching wrong things, and making the world blind in process management technology.

This is the collection of many forms of communication with those who were in contact with me to answer their questions, comments on my papers written in LinkedIn, YouTube, and other media. Also, you may see the published articles from various industrial magazines, especially on process control techniques used, in particular - Six Sigma, Black Belt, SPC, Cp and Cpk etc. My experience has proved that the statistical application in a manufacturing set up is not the right way to control the process. For some way or other, those who published articles and presentations on various forms of statistical tools and techniques, have made their fame and money, but for the consumers those techniques made truly little improvement, or nothing accomplished at all.

Those techniques were way old when the manufacturing technology was in primitive stage. Time has changed, technology has changed several fold and changes are taking place continuously to make the process more visible and transparent.

I was very much interested to learn more about the statistical applications in a manufacturing set up in the middle of 1970s. When I learned more about this technique, attending several hours of training, attending seminars, taking classroom training, and obtaining my Green Belt program from a prestigious university, I found this knowledge is not applicable to a professional like me, where my core knowledge is in engineering, manufacturing, supplier management and communication with transparency.

I obtained my US patent on "Solid Model of Statistical Process Control" and the details are available for review at http://patft.uspto.gov/netahtml/PTO/srchnum.htm

On the query, in the space, type this number: 7,457,679 B2 and click "Search".

That will take you to my US patent.

"Solid Model of Statistical Process Control"

R SPC is the software created from this model, that is easy to use to plan and address all the risks down the line ahead as a warning and take necessary actions before that damages the product under production.

When I created an engineering solid model of the famous bell curve, it amazed me to see that opened up the door for valuable information as far as the process control in a proactive mode. That means this model helps to define the process control attributes without any data, that means tremendous cost savings to users of the program and the users will be able to make hardware 99.99% in compliance to the quality requirements and the results from the sample run are given in this book.

Teaching wrong things to the students or to the working public is wrong, even if it is through universities, colleges, or through any other media, it is fair cheating, this should be stopped through legislation. Teach what is valuable to the manufacturing industries and to the community.

Industrial Statistics and Commercial /Business Statistics

This is my prediction based on the observations of COVID-19

This COVID-19 has paralyzed the industries by shutting down all the manufacturing industries throughout the world, that made the businesses shut down, that made the economy go down, schools are shut down for an indefinite period, education from home has killed the hands-on laboratory or workshop at the university and college level and made no knowledge able to be gained from education.

This post COVID-19 era is going to be a total transformation for all the manufacturing industries and the human lifecycle. Things are not to be the same as before. This is the best time to retool or reform or reorganize the operating system with knowledge induced rather than statistically polished. This is the best time for the statisticians to jump into the field to collect data from all over the world and do research and come up with a medicine for this sickness, the best contribution they can make to the world.

Sow the seeds of quality and reap the harvest of quality products for a reduced cost.

1. Revise your system procedure to make it simple, you may already have a system, just revise to close the gap.

2. Create work instructions that controls your manufacturing processes and techniques, this is your confidential document, nobody has the permission to review this. (We will help on this)
3. Instead of shop traveler, create process maps. (We will help you on this)
4. Control the variables in a manufacturing process, process map will help you to do that. (New tool)
5. Do not spend time and money on doing SPC (traditional statistical process control) to measure the goodness of the product, it is a waste of time and money, you know it by now.
6. Go for Reverse SPC. It is easy and simple and helps to control critical parameters in a process without any additional cost. (New tool) Try and see the values.
7. Supplier management and controls are extremely easy and transparent. (We will help you on this.)
8. Employee training and customer interaction. (We provide training)
9. Go for second party audit and if they certify and approves your process and system, why waste money and time on third party audit? (We will help you to communicate with your customers if needed.) Experience has proved that the third-party auditors are mostly less product and process knowledgeable. That is why different auditors come up with different observations and recommendations. The best fit is letting the customer do the system audit and approve the system, they know their product and processes. That is called second party audit.
10. These are the fundamental changes you have to make, to run your operation smoothly and efficiently and all the problems will go away, with a cost-efficient, proactive approach to win the manufacturing and quality competition of the future.

When you read this comparison chart below, you will wonder and shocked that the amount of money being wasted all these years, kind of driving the dead horse.

Statistics were in existence several thousand years ago from the existence of the human being on earth. It is a tool to collect data and make the decision whether to accept or reject decisions. Predictions are made based on the data collected over the years. Late quality guru Dr. Edward Deming said repeatedly that yesterday's data will not predict the future and he stressed the importance of process and product knowledge for all those involved in the manufacturing sector. Probability, reliability, and statistics all these theories have a common built-in deficiency, that is – uncertainty or lacking confidence or error to a certain percentage based on the accuracy of data. In a way the SPC is a high-level bargaining tool to sell the bad products to the customer.

Difference between Traditional SPC and Practical Industrial Statistics (R SPC)

SPC	R – SPC
1. Data dependent	No data (old) required.
2. After-the-fact process	Before the fact
3. Visibility is narrow	Full transparency/visibility.
4. Evaluates one variable at a time	Evaluates and control more than one variable.
5. Focuses on critical variables	Consider all variables are critical.
6. Operating cost is high	Cost impact is very small
7. This is a separate process	This is built into the manufacturing process.
8. No process and product knowledge required	Requires process and product knowledge.
9. Requires several parts to collect data	It follows artificial intelligence model.
10. Not good for small production run	Very good tool for expensive small lot.
11. Quality of the products not guaranteed	Product quality is guaranteed.
12. Risk factor is high to customer (product)	Risk factor is very minimum or nil.
13. This is an age-old technique	New, tool a game changer, process management.
14. Supplier control is difficult	A companion to the supplier's control.
15. Remote monitoring not possible	Remote monitoring is possible.
16. Paper documentation runs high	Paper consumption can be minimized.
17. Return on investment is not guaranteed	Guaranteed return on investment.
18. Communication is very minimum	Communicates with all levels of people.

19. Handling inspection damages the product	Reduced inspection and handling of product.
20. Customer satisfaction low	Higher customer satisfaction
21. Politically motivated and pushed	No room for politics, it is real.
22. Bad parts end up for customer review	Error proof and elimination of risk.
23. Accepts bad products and rejects good	All products are made with the same quality.
24. Slows down the production	Accelerates the production.
25. Industries are looking for a better way	This is the tool for the future of industries.

We provide training and implementation of this tool to the customers; we will walk them through the process until the products are ready for the customer to use. We have no intention of selling the tool by itself without proper training and education. It is a guaranteed return on investment to those who invest in this program within a year or two. We provide technical consultation service to potential customers for a fee assessed separately as needed.

Scripture says (Matthew 13:8) "But the seed fell on rich soil, and produced fruit, a hundred or sixty or thirty-fold."

It is your money, and you spend it in a way you wanted it or if you want to spend the money wisely to work the investment for you to generate profit, you just change the course of your journey in the right direction. This is called industrial transformation for the twenty-first century. Sometimes the treasure is buried in the sand, and we have that treasure of knowledge for you.

Please check https://www.qem-inc.com and contact us for more details.

INDUSTRIAL STATISTICS. NEW GAME CHANGER FOR THE MANUFACTURING INDUSTRIES.

I started a quality engineering management consulting business with the idea of giving training and a renewed education to the millions of manufacturing industries across the globe. I know that the saying goes like this "it's hard to train an old dog new tricks" and it is true, but I am trying my level best to bring some light to those who are blindfolded by fake education and training by those whose intensions are only to make money and fame for them, not to those who are taking their training course.

Bob Matthew: Process approach is my strength that gave wonderful results of higher quality products, timely delivery, reduced cost, and reduced waste and above all happiest customers.

Process approach means giving more emphasis on process and tailor the process map according to your product. In the process approach scenario, process control is the key.

These are the additional instructions on how to use the R SPC program, it is a Windows-based program that can be downloaded to your computer for a nominal fee. Upon paid and confirmed receipt of money, you will be given authorization to download the program to one single user. If you need more, each computer must pay individually.

Visit: https://www.qem-inc.com for more details, how to buy this program for a single terminal user.

Use this instruction in conjunction with the Help File User (R SPC)
Before you use this program, you must do the following:

a. Read the customer contract and note down all the deliverables and the time of delivery.
b. Read the engineering drawing or engineering model to understand what is asked to make by the contract. Use ASME-Y-14.5M standard 1994 version as your master guide to review the engineering drawing to assure all the variables and notations are properly addressed in the drawing or engineering model. Make corrections to the drawing or model before you proceed further, simple reason is that a wrong drawing would not make a quality product.
c. Review the engineering drawing and serialize all the variables. This includes the engineering notes or special instructions. This process is particularly important for the accuracy and compliance with this program. All these are going to be part of the deliverables.
d. Say for example: The drawing number is 12345 and this drawing has ten different features and number them as – 12345-1, 12345-2 —- and 12345-10 etc.

Note: This is powerful program that helps the manufacturer to fabricate every component of a major assembly or a simple product or part of a sub assembly or major assembly with confidence to produce 99.99% quality products with reduced cost, lean manufacturing and time saving. This program eliminates the extensive use of SPC that will not tell you anything about a process or quality of the manufactured part. This tool gives you cost savings from the beginning of the use of this program. Using SPC, that analyze only one variable, but this RSPC program defines and controls all deliverables in a contract.

[Caution: It is a US patented and copyrighted material; therefore, copying it in any form or in a different media or format or in a different language, create a software in a different manner, it is prohibited. Get the authors permission before doing such a mistake.]

Key element of the R SPC software program.

Understand that there are many forms of dimensional variables in an engineering drawing or an engineering model, examples are given below:

Dimension. Example: Unilateral, Bilateral, Range and True position and understand that there are five types of tolerances, and they are: From, Profile, Orientation, Location and Runout. You should master this terminology and its use and applications. Refer ASME-Y-14.5M, 1994 standard and learn the type of dimensions and tolerances, these are the fundamentals you should learn it to perform excellence.

An inhouse training is suggested for speedy understanding of the program and measure the advantages, cost savings and quality measure at the same time, most importantly you and your staff will gain confidence and trust in this program.

Contact: 714-783-8911 for more information.

R SPC Screen Shots are given below with explanation for clarity.

R SPC Software

This program has the main screen: Home, Client, User, User Logs and Logout.

Below the main screen, there is a sub heading Sample List,

Below that: Add New Sample, Compare Bell Curve, CSV, and Excel.

NOTE: Refer to the User Guide talking about the master data. This master data is created by the program based on the input for each variable that you identified with serial numbers and put into the program. [When SPC controls only one variable for analysis, this RSPC program controls all variables in a process without any additional cost]

After you complete input all the variables into the program, go to the Search features and type the drawing number of the part that you just finished input. Search function will display all the dash numbers for that part. Click the master data, the system will display either in CSV format or in Excel format. Save this data and you are going to work with this data for programing your NC, CNC machines.

Look at the master data that you saved in your computer in the Excel format or CSV format. Let us see the Excel format and you may see data in Column – A through AG. This is data you input into the program. You may notice that the variables you input have standardized by the program and created Derived Dimension and Derived Tolerance in Plus or minus format, without changing the design intent, and we are going to see the data in Column S through Z in detail and this is the key element of the program.

The bottom chart shows:

W = Derived dimension from master Data, this is mean dimension.
S = Derived standard deviation – that equate to capability of the machine.
V, U, T = One derived standard deviation deducted from value under W each.
X, Y, Z = One derived standard deviation added to the value under W each.

S	T	U	V	W	X	Y	Z
Standard Deviation	X-3σ	X-2σ	X-1σ	X	X+1σ	X+2σ	X+3σ
0.0013	0.8751	0.8764	0.8777	0.8790	0.8803	0.8816	0.8829
0.0017	0.9699	0.9716	0.9733	0.9750	0.9767	0.9784	0.9801
0.0018	0.7446	0.7464	0.7482	0.7500	0.7518	0.7536	0.7554
0.0008	0.3751	0.3759	0.3767	0.3775	0.3783	0.3791	0.3799
0.0010	0.7440	0.7450	0.7460	0.7470	0.7480	0.7490	0.7500
0.0008	0.9751	0.9759	0.9767	0.9775	0.9783	0.9791	0.9799
0.0012	0.8729	0.8741	0.8753	0.8765	0.8777	0.8789	0.8801
0.0020	0.9750	0.9770	0.9790	0.9810	0.9830	0.9850	0.9870
0.0017	0.8699	0.8716	0.8733	0.8750	0.8767	0.8784	0.8801
0.0250	1.2500	1.2750	1.3000	1.3250	1.3500	1.3750	1.4000
0.0005	0.6485	0.6490	0.6495	0.6500	0.6505	0.6510	0.6515
0.0013	18.0001	18.0014	18.0027	18.0040	18.0053	18.0066	18.0079
0.0025	0.9840	0.9865	0.9890	0.9915	0.9940	0.9965	0.9990

Please remember, statistical process control application starts only after making the products and they focus on only one variable. Industrial statistics focuses on all the variables of a product. Additionally, manufacturer has the advantage to select dimensions from the master data S through Z, like shown above, based on the secondary process such as Plating, Hard anodizing, Heat treats and other processes. Another advantage with this Program is the manufacturer can select the variables according to the MMC and or

LMC. Example shown above red circled dimensions are the ones selected to program the CNC or NC machines. Selecting the dimensions from the table above are correct and never violate the drawing dimensions, for processing purpose we intelligently plan and control the manufacturing process to get the desired quality yield. Keep one thing in mind that the machine's capability you are choosing should meet the standard deviation as per the table above column S.

Standard deviation = derived standard deviation, computer generated standard deviation. This standard deviation signifies the capability of the machine to hold that dimension to the selected target. It is critical to maintain the capability of the machine to the manufacturers standard and a performance card should be posted on each machine their capabilities with date and time verified those capabilities. Inspection data by the operator would be good enough to determine the capability of each machine in each axis of the machine. CNC uses this data to select the machine as appropriate to the parts being made by the machine. Standard deviation under column S is the theoretical standard deviation determined by the RSPC program.

Selecting the target is based on the function of the part or to machine the part at MMC. MMC is the maximum material condition of the feature.

Say the part in question going to be machined with a single CNC machine that has a capability of 0.000" to 0.0010". Look at the column Standard Deviation, you may see the capability of the machines can be from 0.0005" through 0.0250". In this case to get the best performance we take machine with 0.0005" capability.

Master data of the R SPC program creates a table of tolerances for each variable you input into the program in the initial stage. Core idea of this program is to give a tolerance band of dimensions from X-3 through X+3 , this allows the CNC programmer to select each variable according to their criticality in the next operation such as: plating, Anodizing or similar secondary processes.

The demonstration shows here in the chart above with red oblong circles, those are the dimensions picked by the CNC programmer to program the machine and all other setup and part holding techniques are done as usual for the preparation for machining the part.

The other part of the process control is also very critical, that is the tooling, fixtures, coolants used, chip removal, mounting the part onto the machining fixture, correct NC or CNC program also considered as part of the tooling, speed, and feed of the machining process, follow the machine manufacturers recommended feed and speeds to get optimum results and long life to the machine. Keeping a scheduled maintenance program for all the machines will extend the working life of the machine. It is recommended to identify each machine with a tag to show the name of the machine, calibration date, expiry date, process capability in each axis, part inspection data will help you to identify these features.

The idea behind this R SPC Program is:

1. To standardize all the variables or dimensions into one format.
2. The master data shows the dimensions in the increments one theoretical standard deviation from X-3σ through X+3σ.
3. It is very important that the capability of a machine should be measured and selected according to their capability. Here we chose machine with capability or standard deviation 0.0005"
4. If the instructions are followed correctly before machining, none of the dimensions will go out of tolerance after machining and the part will be good.
5. Operator can measure part comes out of the machine and input into the sample data of the program, or you may use the automatic measuring and data collection feature of the machine, see page 7 and 8 of the User Guide.
6. Once the data are input into the sample data, program will display the X-bar and R chart along with a bell curve that shows where the data are concentrated. Refer to page 10 of the User Guide.
7. When the work is completed, there is no need of doing another data collection to do SPC to verify if the process is/was in control and check the quality of the parts made by this program. This R SPC program has already taken care of that process that assures the quality of the part(s) made are good.

A. This is the best alternate process control method where the processes are defined, and control parameters are specified before beginning the real manufacturing process. It is a total transformation to the quality management process to speed up the production of parts and assemblies to deliver to the customer at the earliest. No need for a quality person to look over your shoulder to assure you are doing OK or otherwise. This process creates more confidence in the management, employees, and the machine operators. Process visibility is the key.

B. Review the engineering drawing or model to assures the compliance to the GD&T requirements. A bad engineering design will not make a quality product. So, the prevention starts from here. Review of the drawing should be done by a professional before even using the R SPC program to input all the variables in.

C. Process mapping or an advanced shop traveler helps to identify any risks down the line while making the part in planning, this is an added feature of the R SPC program.

D. Process mapping identifies the kind of measuring instruments are supposed to use during the fabrication of the part with tool ID number. Qualify the operator to use this tool.

E. An unprofessional shop traveler can create problems during the fabrication of the part. A well-planned process mapping can eliminate risks in advance.

F. Cost of MRB activities, rejection and rework, customer return and many other costs are not reported and that stays within each company. But we do know that there have been several incidents of recalling the products already on the market, that indicates that there were problems with the company maintaining and delivering quality products.

G. Since the technology is changing rapidly in the manufacturing sector, quality organizations are not keeping up to align with the technological changes in the design and manufacturing sector.

H. If the customer demands newer products at the earliest, our current technology does not meet the customer's demand. The reason is quite simple, our system is very antiquated, not capable of handling the customer demand.

I. Best way is to get rid of ISO certification process, third party audit, statistical application to check the quality of the manufactured products etc. Suggests going for second party audit and certification, it is the fastest and highly effective and cost-effective way to do business with the customers. ISO has its own problems to stabilize the standard, designers or this ISO standard are spinning their wheel to get a good statistical balance of the standard. Go for our old MIL-Q-9858A quality program requirements as a model and create quality manual and work instructions accordingly. Easy to manage and control the process if you knowhow to do it efficiently.

R SPC Program has a Search function that can be used to search by any features like: Customer, Project, Drawing/Part #, Creation Date, Revision date, Revision Code and Process #

There are different windows to specify the details about the Customer, Project, Drawing Part Number, and Creation date (of this ADMIN_#) Revision date, Revision code and Process #. These are the mandatory fields it should be completed, and these features can be searched. There are optional fields, intended for specifying more details of the part for traceability and identification purpose. After filling all the required fields, press the save button to save the data entered.

Next window to identify the type of dimension you are processing and there are Three types you can choose from: they are: Linear, Range and True Position.

Some of the screen shots from the R SPC program is given for your reference only. Make sure to read and understand the User Manual before you start using the R SPC software.

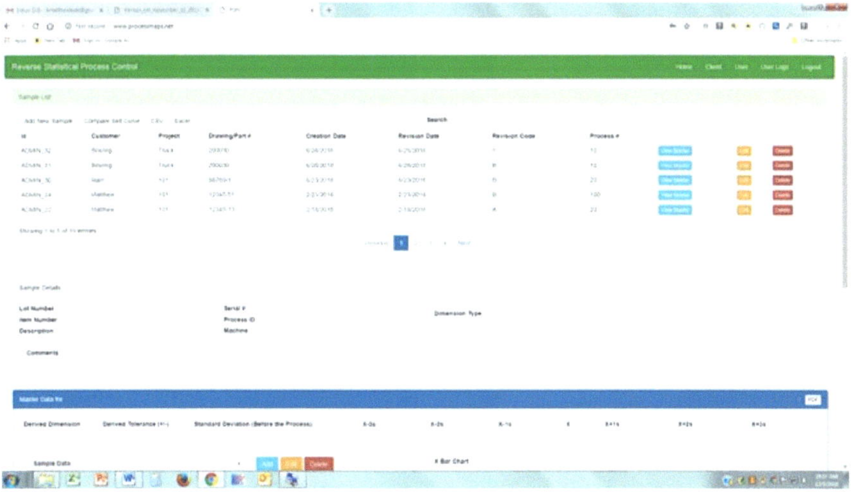

Sample Details are shown below:

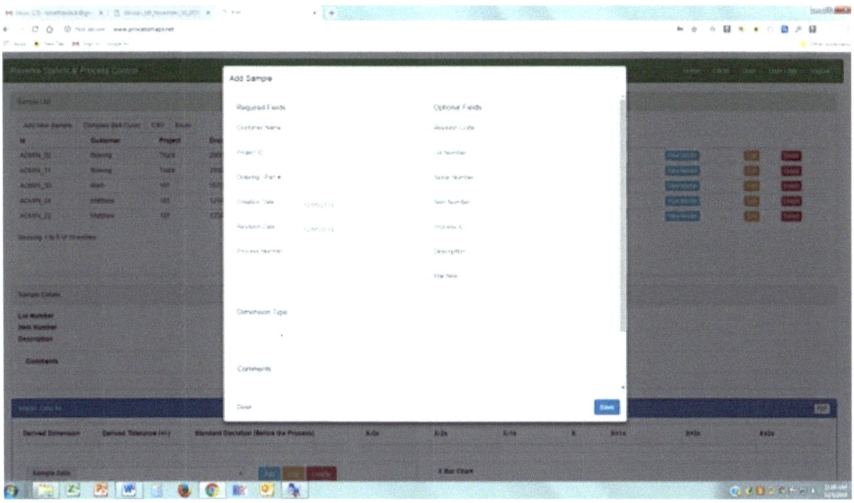

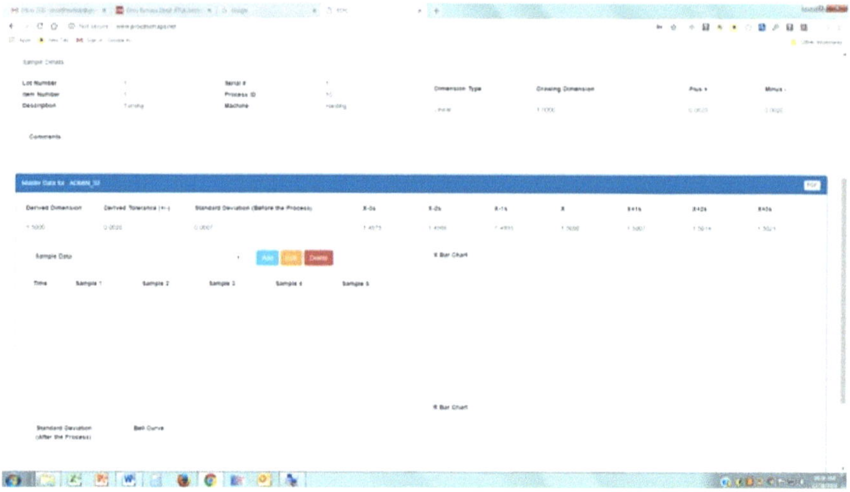

R SPC screen shot bottom part.

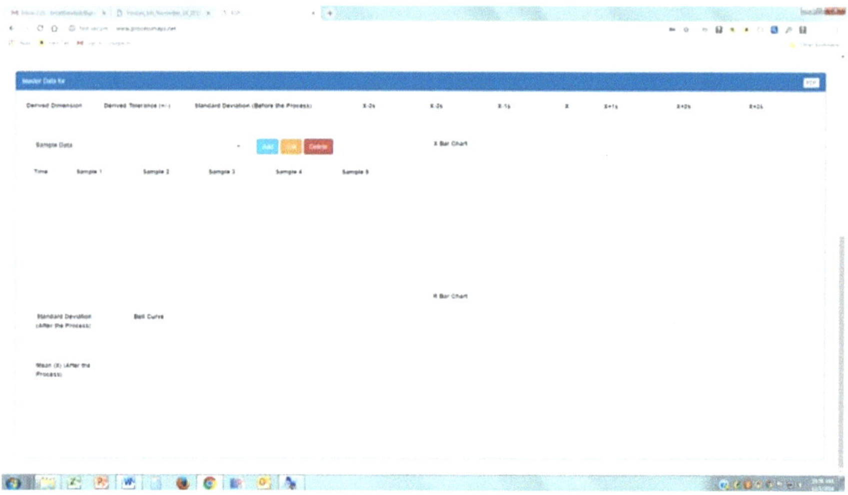

Add Sample

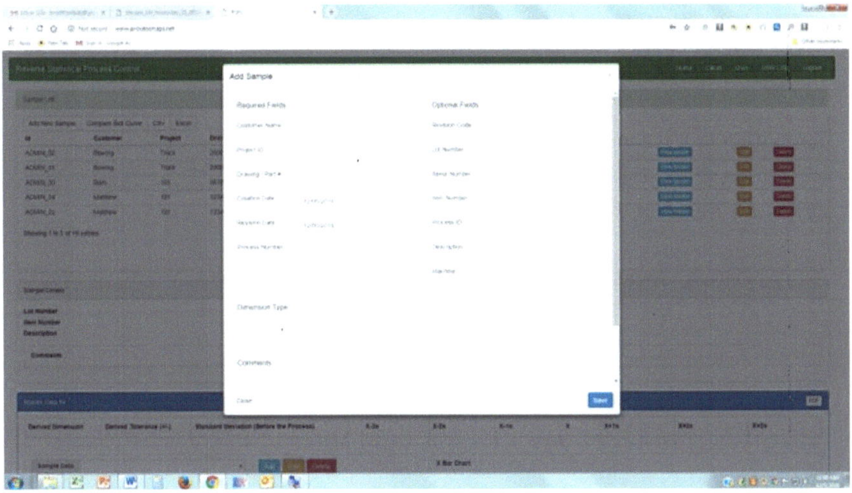

Sample Details

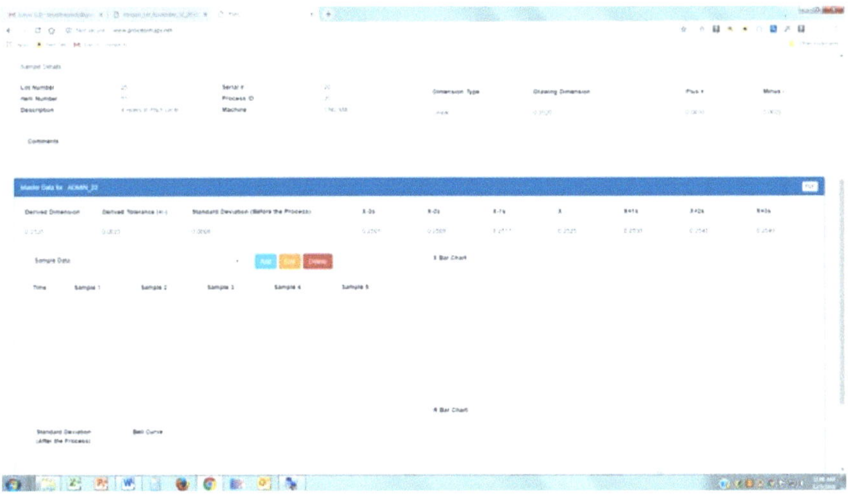

Master Data for Admin 22

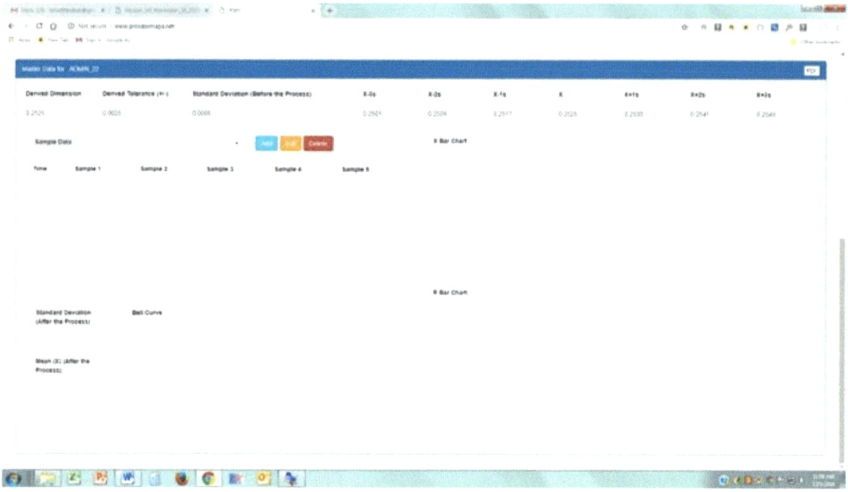

Sample Data for Admin 22

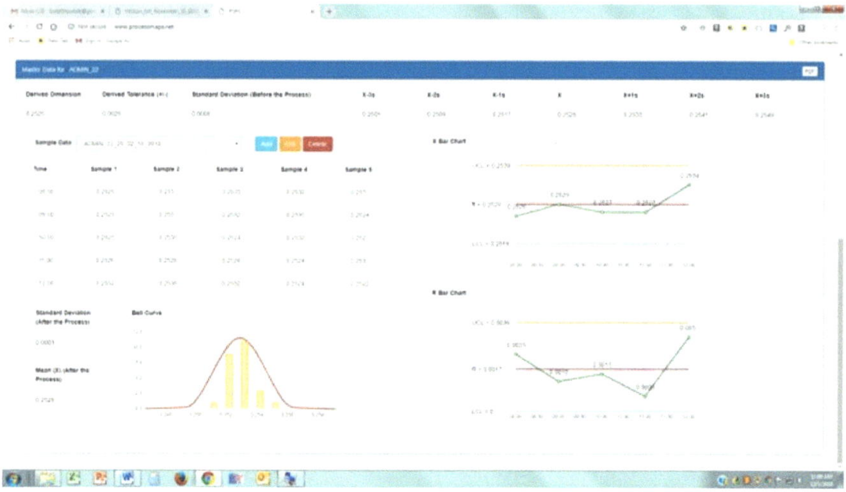

Compare Bell Curve(s)

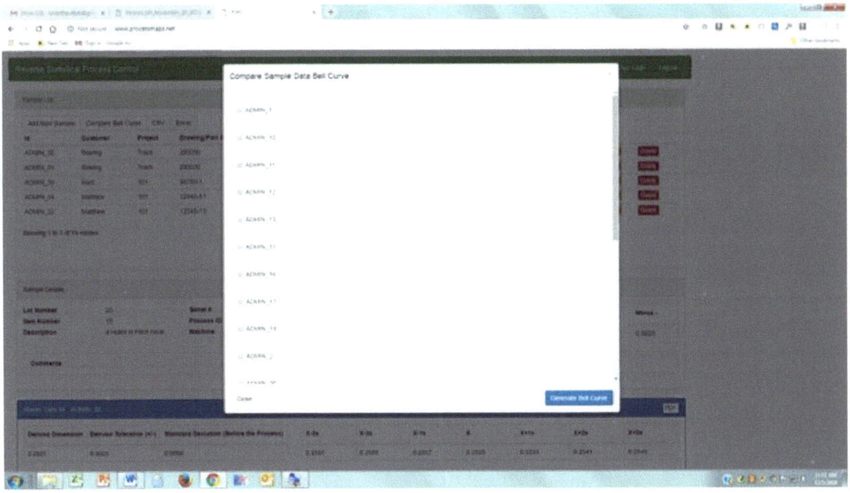

Choose/ Pick the Bell Curves to review.

Choose – Generate Bell Curves Picked (5)

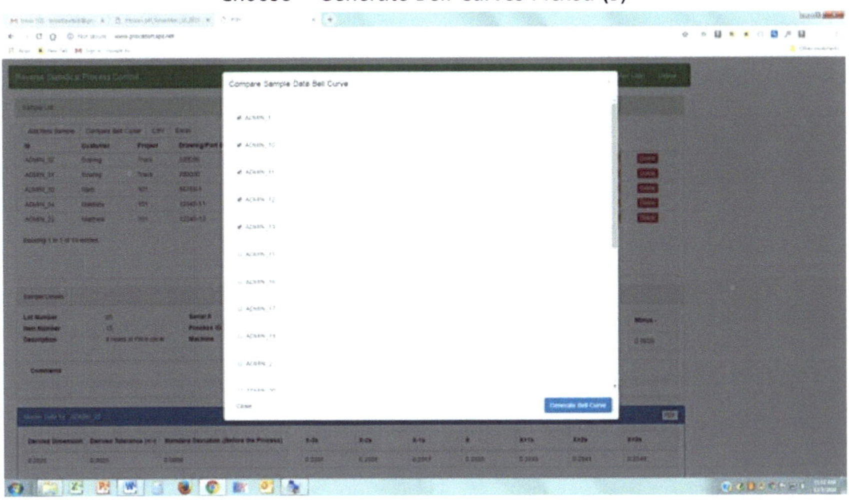

Go to https://www.qem-inc.com If you want to access the web based software give your full name, address, company details to access the program. Upon receipt of the payment, we will send you a code to access the program for one terminal user only. Procedure is the same for multiple users in the same company.

END OF DISCUSSION ON, RSPC, INDUSTRIAL STATISTICS AND TRANSFORMATIONAL QUALITY MANAGEMENT SYSTEM

Choose – Generate Bell Curves Picked (5)

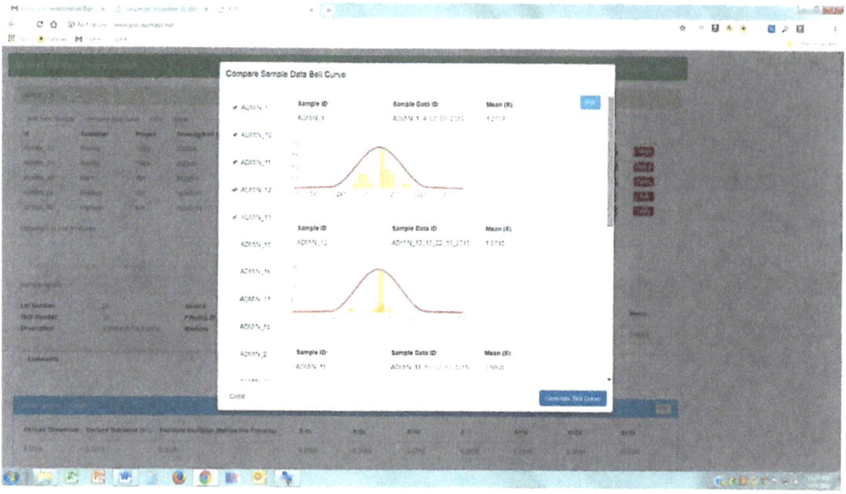

INDUSTRIAL STATISTICS.

BIRTH OF A NEW TECHNOLOGY

See the statement and working principles of industrial statistics and transformational quality management system. You may see some of the presentation materials posted here from the PDF files.

INDUSTRIAL STATISTICS – NEW APPROACH

WHAT IS INDUSTRIAL STATISTICS?

1. It is the statistics about the industrial manufacturing machines
   a. Talk about: Lathes, Mills, Gantries and Grinders etc.

2. How they are designed and manufactured
   a. Each manufacturers have their own standards for design and its functionality (Three axes, five axes, automatic, CNC etc.)

3. Manufacturer's focus for each machine's ability
   a. Ability of each machine is different like, milling, turning, threading etc.
   b. Precision and accuracy are different like, hold tolerances to 0.001", 0.0005", 0.0001" etc. This may vary with load and without load. With load means, maintains the same accuracy during operations in removing or cutting the metals.
   c. Repeatability is the capability of a machine, is the ability to sustain the accuracy/precision at normal temperature and pressure during the life of the machine.

 1

QUALITY MANAGEMENT SYSTEM

Why Industrial Statistics is required now?

Our manufacturing industries were / are going through several roadblocks, because of the roadblocks, industries could not perform their maximum potential in making parts and assemblies to support our industries, such as Aerospace, Automotive & Medical device manufacturers.

Biggest roadblocks; Quality management system and their strict requirements, based on ISO, GMP & AS9100 and so on.

Next and the biggest enemy is the Statistical Quality Control and Statistical Process Control, Six Sigma, Lean Six Sigma, Cp, Cpk etc. and they do not know the meaning of Sigma

 2

QUALITY MANAGEMENT SYSTEM

What is wrong with our Quality Management System?

1. Our quality management system is a failure.
2. It is a compartmental system, less communication and imposing the requirements those are hard to comply with and has nothing to do with product quality at all.
3. Just audit the process and write up for noncompliance, a process that ties up the hands and feet of manufacturer.
4. Third party audit and certification, very politically influenced process, most of the auditors have no process knowledge the area they are auditing
5. Use of PVA (Process Validation Assessment) checklist, checklist created by the professionals who knows the process, but those who are using it are not knowledgeable in the process being validated.

QUALITY MANAGEMENT SYSTEM

What is wrong with the Quality Management System?

6. To eliminate the above discrepancies, introduce a new Transformational quality management system, that is based on the model MIL-Q-9858, an old efficient TQM system, it is easy, less expensive and connects throughout the system.
7. Quality system is like a fully organized Orchestra, in that everyone is instructed to do (Play their part) their part according to the instruction provided earlier.
8. If everything is fine tuned and given all necessary, check and balance methods and tools, the process owner would be able to program the machine and make parts correctly to the specification.

MANUFACTURING PROCESS CONTROL

Contributors of efficient manufacturing process involves: Man, Machine, Material, Methods, Measurements & Environment.

1. In that Man is the key person or called the process owner
2. Good technical knowledge of the machine
3. Know the capability of the machine, see manufacturers Spec.
4. Process capability of each machine should be displayed on each machine with the last date of verification, it should be done at least once a year, it is recommended to revise it once in six months.
5. Please remember that the capability of the machine is measured as the capability of the machine with load. That is measured by checking the part(s) at a certain interval by the operator and document it. That data is the capability of the machine in real time.

 5

STANDARD DEVIATION – σ

Let us discuss about Standard Deviation.

1. There are two types of Standard Deviations. The most important Standard Deviation in the manufacturing process is the capability of the Machine, we discussed before, this is the real time Standard Deviation.
2. In the R SPC software program, this real time Standard deviations are used to define and control the manufacturing process. There will be more discussion this later.
3. There is a vast difference between the Statistical Standard Deviation and Real time Standard Deviation.
4. Some discussion on Statistical Standard Deviation are presented in the following slides, remember that the Statistical Standard Deviation is an after the fact measure.

 6

STANDARD DEVIATION – σ

In Statistics - Area Under Normal Curve

Area under normal curve. Why it is called normal curve?. According to the Empirical Rule or 68-95-99 Rule, In statistics it is assumed that the process is under control & normal. In another words it is a "Cookie Cutter Process Application"

Then $Z = \frac{X - \bar{X}}{\sigma}$

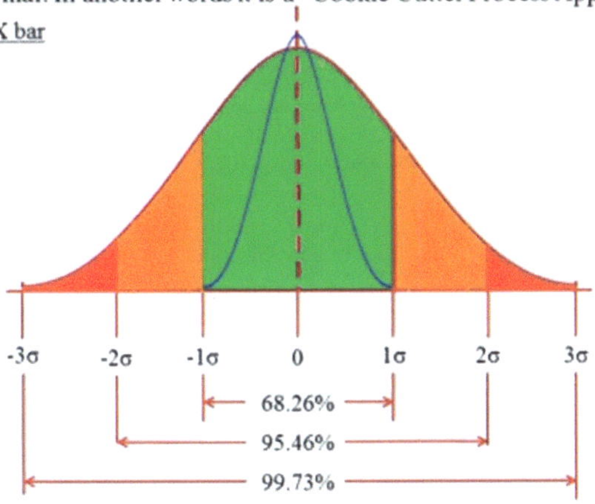

STANDARD DEVIATION – σ

Standard deviation equation for an entire population:

$$\sigma = \sqrt{\frac{\sum(x - \overline{x})^2}{N}}$$

where

$\sigma =$ *the standard deviation*

$x =$ *each value in the population*

$\overline{x} =$ *the mean of the values*

$N =$ *the number of values (the population)*

$\sigma^2 =$ Variance

 8

STANDARD DEVIATION – σ

Standard Deviation. Basic Examples

Consider a population consisting of the eight values: 2,4,4,4,5,5,7,9

Average of this eight data points = 2+4+4+4+5+5+7+9 = 40/8 = 5

Mean or Average = 5

To calculate the population standard deviation, first compute the difference of each data point from the mean, and square the result of each:

$(2-5)^2 = (-3)^2 = 9$ $(4-5)^2 = (-1)^2 = 1$ $(4-5)^2 = (-1)^2 = 1$

$(4-5)^2 = (-1)^2 = 1$ $(5-5)^2 = (0)^2 = 0$ $(5-5)^2 = (0)^2 = 0$

$(7-5)^2 = (2)^2 = 4$ $(9-5)^2 = (4)^2 = 16$

Next, compute the average of these values and take the square root

$\sqrt{(9+1+1+1+0+0+4+16)} = \sqrt{32/8} = \sqrt{4} = 2$

Therefore, Mean or Average = 5

And Standard deviation of Population = 2

$\sqrt{(9+1+1+1+0+0+4+16)} = \sqrt{32/(8-1)} = \sqrt{32/7} = \sqrt{4.571429} = 2.138$

Standard deviation of Samples = 2.138

 9

STANDARD DEVIATION – σ

STANDARD DEVIATION IN STATISTICS

$$S = \sqrt{\frac{\Sigma(X-\overline{X})^2}{N}}$$

where S = the standard deviation of a sample.
Σ means "sum of"
X = each value in the data set
$\overline{X}$ = mean of all values in the data set
N = number of values in the data set

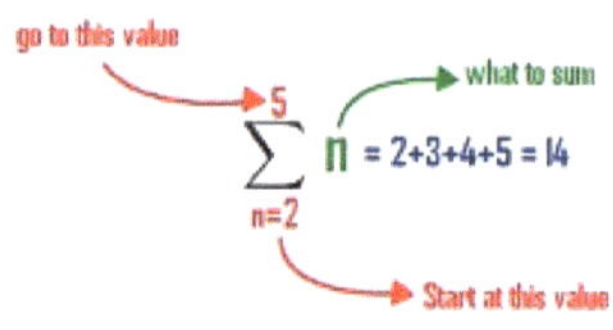

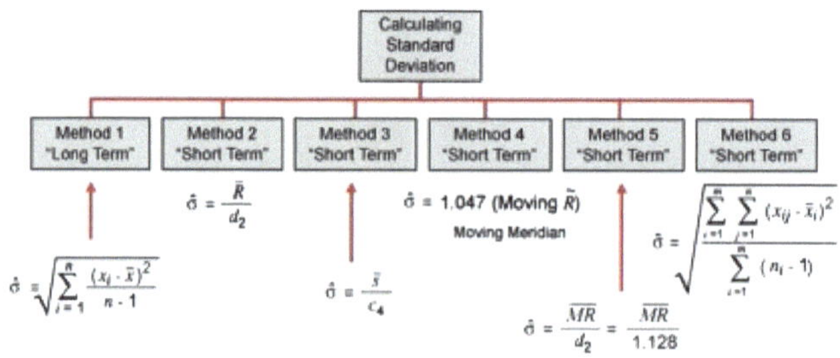

 10

STANDARD DEVIATION – σ

Typical Application of Statistics

In a precision manufacturing process, the diameter of Steel Shaft is normally distributed with the mean of 0.2508", and the Standard deviation is 0.0005. If the specification of the Shaft is 0.2500 + / - 0.0015", what percentage or fraction of the Shafts produced would conform to specification?

Answer = 91.92%

If we could adjust the manufacturing process by adjusting the machine to [0.0008"], so that the process mean is Exactly equal to 0.2500" then the process yield would be 99.73%

According to the R SPC (Real time data) process yield is 100%Min. actual =(0.0005/0.0002) X 100 = 250%. There is no magic here, this is real time.

 11

STANDARD DEVIATION – σ

Statistical Standard deviation is not the right measure to gauge quality of a process and product as many of the business consultants claims Statistical Process Control claims.

Statistics has no room in the manufacturing floor to deal with any of the performance measures in the factory floor. Statistics belongs to some other area besides the manufacturing process, statistics distorts the real image of manufacturing technology.

We are dealing with a process, in that all the elements are defined or definable or everything is laid out or specified. What we must do is to tailor the manufacturing process so that machines controls the variables in each process.

 12

STANDARD DEVIATION – σ

Mathematically defined area under Normal Curve.
In the mathematical model, area under Normal Curve is 100%

Process Yield = $\frac{\text{Calculated } \sigma}{\text{Actual } \varsigma}$ X 100

Ex: (0.001/ 0.001)100 = 100%

Calculated σ = Planned σ
Actual ς = Sample ς

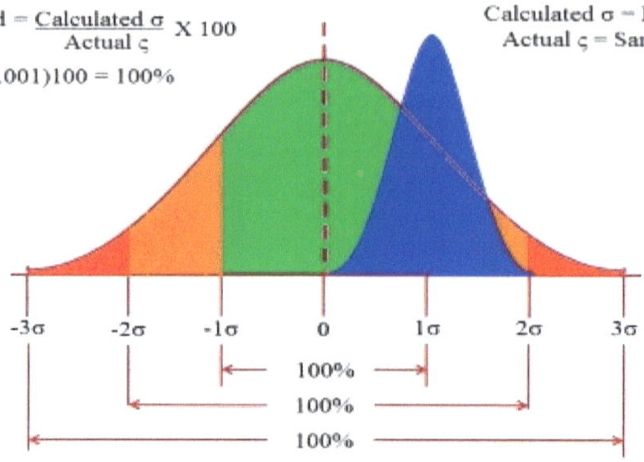

STANDARD DEVIATION – σ

The Power of Strategic Process Control
There is no Magic here, it is real, it is repeatable.

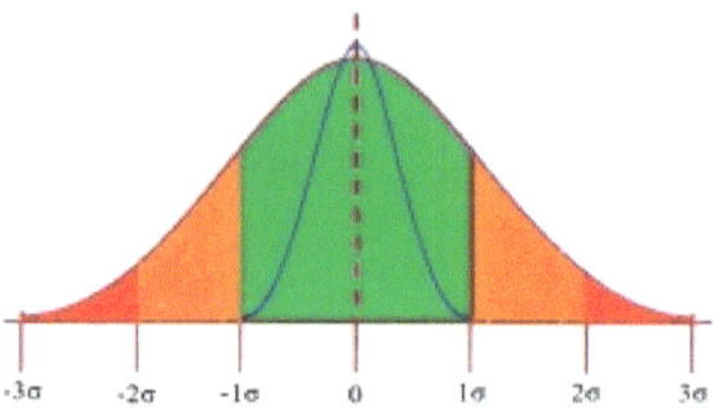

Actual Std Dvn from 25 Samples (Sigma)	Calculated Std Dvn = R-bar/d2. (Sigma)
0.0004	0.0006

Process Yield = $\frac{\text{Calculated } \sigma}{\text{Actual } \varsigma}$ X 100

Process Yield Throughput = (0.0006/0.0004) X 100 = 150%

Is it not better than 6Sigma yield 99.73%?

There is no limit for your cost savings

STANDARD DEVIATION – σ

Statistical Calculation of the Test Results.
We can make this happen at your company.

COMPARISON OF CALCULATED SIGMA (using SPC) AND ACTUAL SIGMA FROM QDC FORM				
CALCULATED SIGMA (using SPC) = R-bar / d_2. Where d_2 = 2.326 For Sub group 5.				
CALCULATED SIGMA				ACTUAL SIGMA FROM QDC FORM
TEST #	R-bar	d_2	σ	σ - Actual
101	0.0014	2.326	**0.0006**	**0.0004**
102	0.0017	2.326	**0.0007**	**0.0001**
103	0.0013	2.326	**0.0006**	**0.0004**
104	0.0044	2.326	**0.0019**	**0.0009**
105	0.0013	2.326	**0.0006**	**0.0004**
106	0.0034	2.326	**0.0015**	**0.0003**
107	0.0007	2.326	**0.0003**	**0.0001**

CORRELATION ANALYSIS

	X	Y
	6	[illegible]
	7	[illegible]
	6	[illegible]
	19	[illegible]
	6	[illegible]
	15	[illegible]
	3	[illegible]
Average	9	4
		1

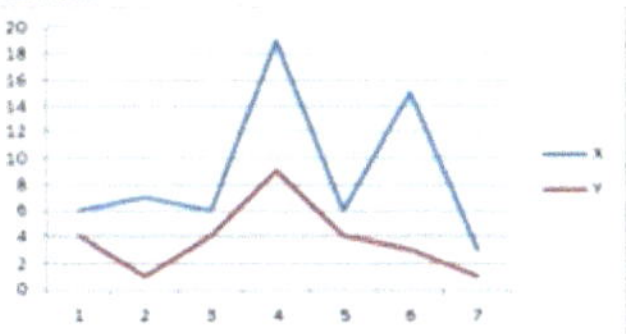

A correlation coefficient of +1 indicates a perfect positive correlation.
As variable X increases, variable Y increases.
As variable X decreases, variable Y decreases.

DESIGN VERIFICATION BEFORE PROCESS

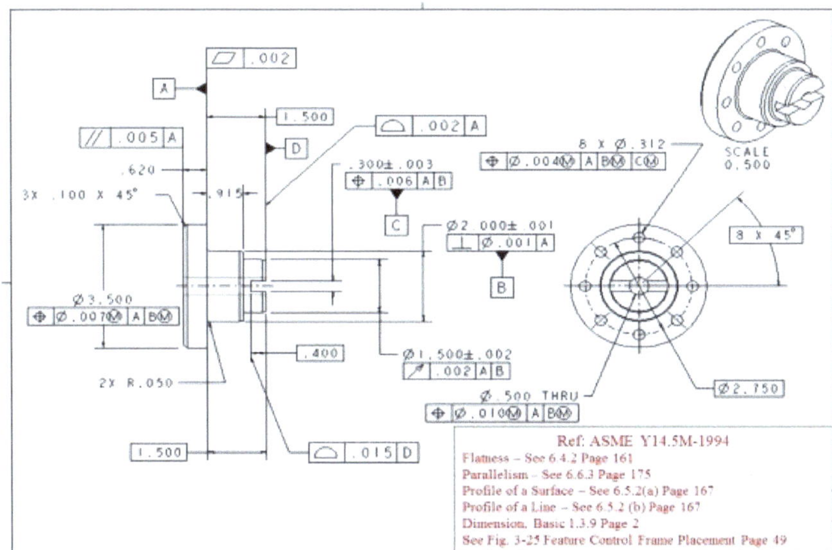

DESIGN VERIFICATION BEFORE PROCESS

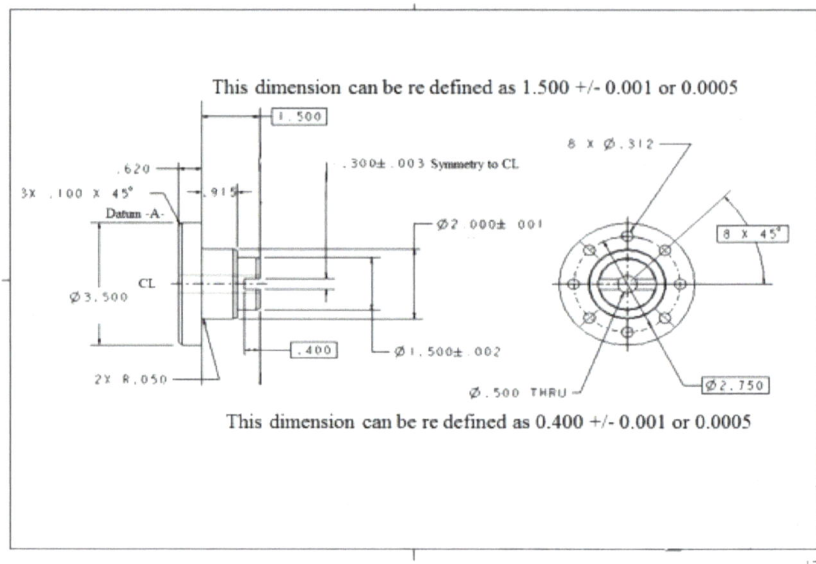

17

PROCESS MAPPING

PROCESS MAPPING (Starts from the component level)

MATRIX

Project =	Project Name	Item number Pt. Number	Item Name	Customer Name	Address	Phone =	Email address
Drg =	Model =	Serial =	Process =	Process Supplier	Purchased from	Approved Processor. Yes No	Machine =
Machine Capability	CNC Program = & Revision =	Machine Stability	Theoretical Std Dvn. From QDC Form	QDC Matrix data	QDC Matrix =	QDC Matrix Revision	Measuring Instrument used
Temperature	Pressure	Humidity	Calibration Control	Employees trained			

Operation =	Description	Control Parameters	Set up proofed	Operator Check	Measuring devices Used with accuracy
10	Turning OD	OD. 0.975 – - 0.002"	1" piece qualified	0.9762"	0 -1" Micrometer, 0.0001"

18

ANALYSIS OF MASTER DATA

Master Data is created by the R SPC software program

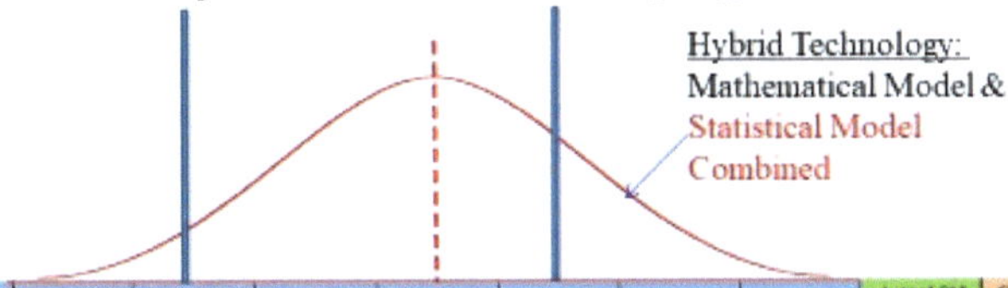

Features	Limits	Machine / Process - Repeatability/Capability/Stability Capability	X-3	X-2	X-1	X	X+1	X+2	X+3	Actual Std Dvn from 25 Samples (Sigma)	Calculated Std Dvn = R-bar/d2. (Sigma)
Linear	Plus/ Minus	**0.0010**	0.6000	0.6800	0.7000	0.7500	0.8000	0.9000	0.9500	**0.001**	**0.0020**

This is the Statistical Model where the data are displayed in +/ - 3 Std. Dvn format.

This is the Mathematical Model where the process is targeted between X-2 & X

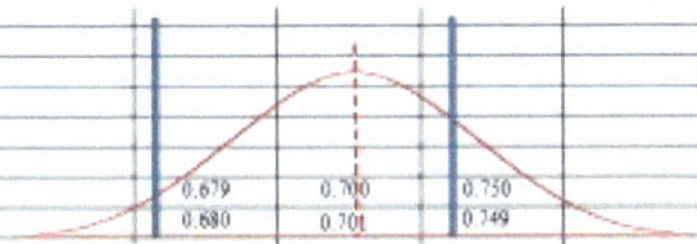

19

ANALYSIS OF MASTER DATA

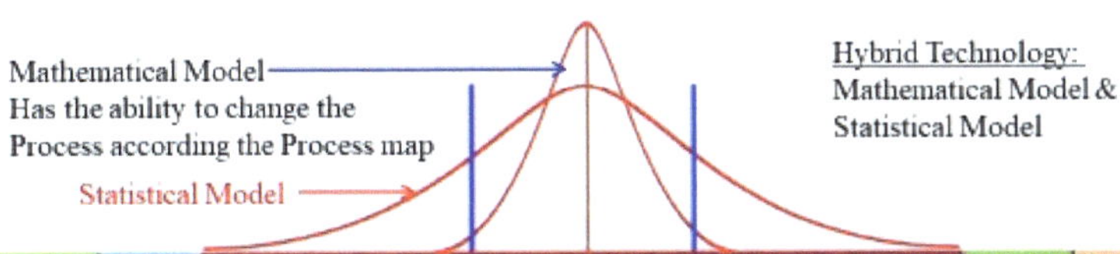

Features	Limits	Machine / Process - Repeatability/Capability/Stability Capability	X-3	X-2	X-1	X	X+1	X+2	X+3	Actual Std Dvn from 25 Samples (Sigma)	Calculated Std Dvn = R-bar/d2. (Sigma)
Linear	Plus/ Minus	**0.0010**	0.6000	0.6800	0.7000	0.7500	0.8000	0.9000	0.9500	**0.001**	**0.0020**

No Artificial Intelligence is needed here.

When the Mathematical model is used, Just use common sense.

20

ANALYSIS OF MASTER DATA

Calculate the Yield.
Process Yield = PC/St. D * 100
(Process Capability/ Standard Deviation)
Process Yield = (0.0010/0.0010)* 100 = 100%

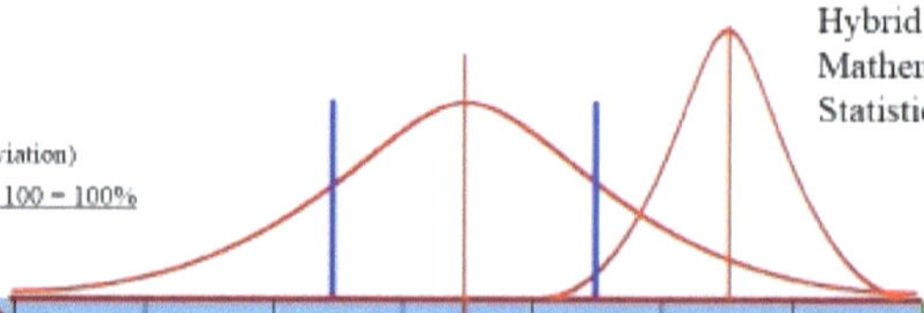

Hybrid Technology:
Mathematical Model &
Statistical Model

Features	Limits	Machine / Process - Repeatability/Capability/Stability Capability	X-3	X-2	X-1	X	X+1	X+2	X+3	Actual Std Dvn from 25 Samples (Sigma)	Calculated Std Dvn = R-bar/d2. (Sigma)
Linear	Plus/ Minus	**0.0010**	0.6000	0.6800	0.7000	0.7500	0.8000	0.9000	0.9500	**0.001**	**0.0020**

21

ANALYSIS OF MASTER DATA

Process capability selected prior to the process.	**COMPARISON OF CALCULATED SIGMA (using SPC) AND ACTUAL SIGMA FROM QDC FORM**							
	CALCULATED SIGMA (using SPC) = R-bar / d_2. Where d_2 = 2.326 For Sub group 5.							
	CALCULATED SIGMA σc				ACTUAL SIGMA FROM QDC FORM			
	TEST #	R-bar	d_2	σ_c		**Min Yield**		Actual σ_a
0.0015	101	0.0014	2.326	**0.0006**		**100%**		**0.0004**
0.0010	102	0.0017	2.326	**0.0007**		**100%**		**0.0001**
0.0021	103	0.0013	2.326	**0.0006**		**100%**		**0.0004**
0.0025	104	0.0044	2.326	**0.0019**		**100%**		**0.0009**
0.0017	105	0.0013	2.326	**0.0006**		**100%**		**0.0004**
0.0020	106	0.0034	2.326	**0.0015**		**100%**		**0.0003**
0.0010	107	0.0007	2.326	**0.0003**		**100%**		**0.0001**
0.0005	112	0.001	2.326	**0.0004**		**100%**		**0.0003**

Process Capability before process ← This is the new Hybrid technology → Use of new technology can boost the quality & productivity significantly.

22

ANALYSIS OF MASTER DATA

Statistical calculation of test results

Process capability selected prior to the process.	COMPARISON OF CALCULATED SIGMA (using SPC) AND ACTUAL SIGMA FROM QDC FORM							
	CALCULATED SIGMA (using SPC) = R-bar / d_2. Where d_2 = 2.326 For Sub group 5.							
	CALCULATED SIGMA σc				ACTUAL SIGMA FROM QDC FORM			
	TEST #	R-bar	d_2	σ_c		MinYield		Actual σ_a
0.0015	101	0.0014	2.326	**0.0006**		**100%**		**0.0004**
0.0010	102	0.0017	2.326	**0.0007**		**100%**		**0.0001**
0.0021	103	0.0013	2.326	**0.0006**		**100%**		**0.0004**
0.0025	104	0.0044	2.326	**0.0019**		**100%**		**0.0009**
0.0017	105	0.0013	2.326	**0.0006**		**100%**		**0.0004**
0.0020	106	0.0034	2.326	**0.0015**		**100%**		**0.0003**
0.0010	107	0.0007	2.326	**0.0003**		**100%**		**0.0001**
0.0005	112	0.001	2.326	**0.0004**		**100%**		**0.0003**

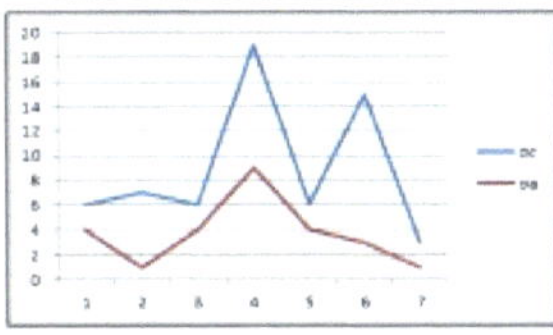

CORRELATION ANALYSIS

	σ_c	σ_a
	6	[illegible]
	7	[illegible]
	6	[illegible]
	19	[illegible]
	6	[illegible]
	15	[illegible]
	3	[illegible]
	4	[illegible]
Average	8	4
CORRELATION		1

A correlation coefficient of +1 indicates a perfect positive correlation.

As variable X increases, variable Y also increases.
As variable X decreases, variable Y also decreases.

Correlation Analysis of the process before and after

23

CHECK OF COMMON SENSE

Water Tank: Purified drinking water

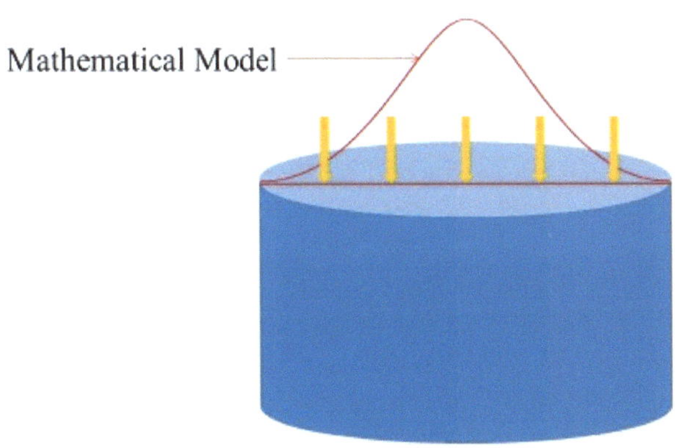

Draw sample from any where from the tank it will be good quality water.

24

DO NOT GET CHEATED BY Cp & CpK

WHY Cp & CpK, ARE THE WRONG CHOICES TO MEASURE THE PROCESS CAPABILITY

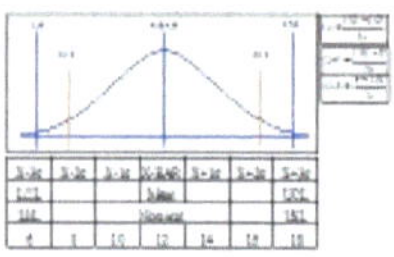

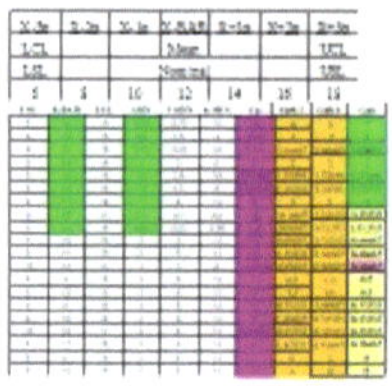

Conclusion on Cp & CpK

1. When the standard deviation decreases, CpK value increases
2. CpK is based on Std. Dvn. & location where the Process setup locked in / value / location
3. This is true only when the process is definable and controllable mathematically.
4. Process become manageable when the process and products are known.
5. If the product and process are unknown, then go for SPC. Collect data, analyze it until you define the datum or baseline of the process. Once the base line or datum has defined then stay away from SPC and do Reverse SPC

 25

DO NOT GET CHEATED BY Cp & CpK

Do not get fooled by Cp & CpK

Model of Cp & CpK. Spec = 12 +/- 6 & $\sigma = 2$

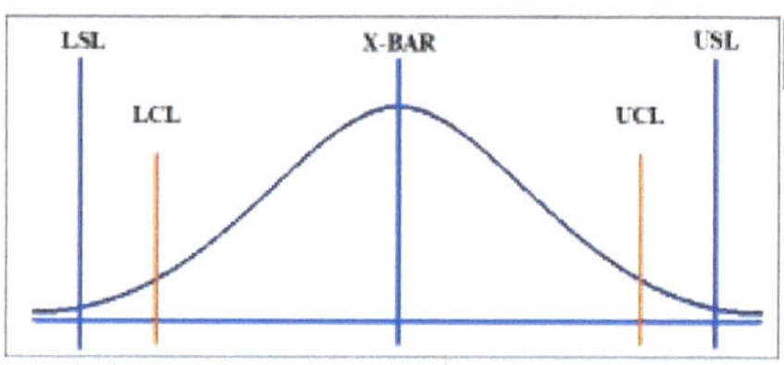

$$Cp = \frac{USL - LSL}{6s}$$

$$CpK = \frac{USL - \bar{x}}{3s}$$

$$CpK = \frac{\bar{x} - LSL}{3s}$$

X-3σ	X-2σ	X-1σ	X-BAR	X+1σ	X+2σ	X+3σ
LCL			Mean			UCL
LSL			Nominal			USL
6	8	10	12	14	16	18

 26

DO NOT GET CHEATED BY Cp & CpK

Specification. 12 +/- 6

X-3σ	X-2σ	X-1σ	X-BAR	X+1σ	X+2σ	X+3σ
LCL			Mean			UCL
LSL		Nominal				USL
6	8	10	12	14	16	18

LSL	X-BAR	USL	SDN	3 SDN	6 SDN	Cp	CpK L	CpK U	CpK
6	12	18	0.25	0.75	1.5	[illegible]	8	8	8
6	12	18	0.5	1.5	3	[illegible]	4	4	4
6	12	18	0.75	2.25	4.5	[illegible]	2.66667	2.66667	2.66667
6	12	18	1	3	6	[illegible]	2	2	2
6	12	18	1.3	3.9	7.8	[illegible]	1.53846	1.53846	1.53846
6	12	18	1.5	4.5	9	[illegible]	1.33333	1.33333	1.33333
6	12	18	2	6	12	[illegible]	1	1	1
6	17	18	0.1	0.3	0.6	[illegible]	36.6667	3.33333	3.33333
6	14	18	0.01	0.03	0.06	[illegible]	266.667	133.333	133.333
6	10	18	2	6	12	[illegible]	0.66667	1.33333	0.66667
6	14	18	2	6	12	[illegible]	1.33333	0.66667	0.66667
10	12	14	1	3	6	[illegible]	0.66667	0.66667	0.66667
6	9	18	2	6	12	[illegible]	0.5	1.5	0.5
6	15	18	2	6	12	[illegible]	1.5	0.5	0.5
6	8	18	2	6	12	[illegible]	0.33333	1.66667	0.33333
6	16	18	2	6	12	[illegible]	1.66667	0.33333	0.33333
10	12	14	2	6	12	[illegible]	0.33333	0.33333	0.33333
6	17	18	2	6	12	[illegible]	1.83333	0.16667	0.16667
6	6	18	2	6	12	[illegible]	0	2	0
6	18	18	2	6	12	[illegible]	2	0	0

 27

DO NOT GET CHEATED BY Cp & CpK

- Do not get fooled by Statisticians
- Do not get trapped by Six Sigma, Lean Six Sigma
- Do not get fooled by Cp & Cpk
  - Conclusion from the experiment shown above
    - As the standard deviation gets smaller , CpK value increases.
    - There is a correlation between Std. Dvn (Machine Capability) and CpK
    - Centering of the process at the Mean has nothing to do with the CpK.
    - It is wise to setup the process to get higher process yield, @ MMC.
    - Process setup should be based on the capability of the machine, unstable machines should not be considered for close tolerance work.
    - When you can work with reality, avoid Statistics
    - Because Statistics is process Blind
    - YOU know the process well; Statisticians do not know the process.
    - Six Sigma, Lean Six Sigma and Statistics may work in the Casinos, Election process, Stock market, Find the source of COVID-19 and the population affected by this sickness and the statistics on life and death data.

 28

BANK WILL NOT COMPROMISE

Bank Loan & Payments

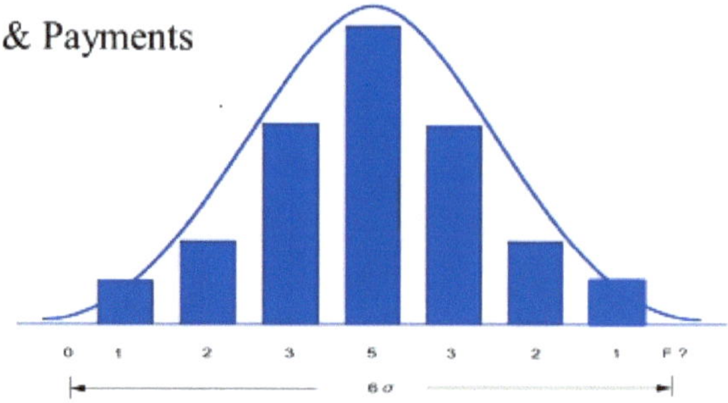

Bank Loan with interest $100,000.00

Made 17 payments of $5500.00 each = $ 93500.00 Balance = $ 6500

One day the loner goes to the Bank Manager & claimed that, look statistically I made all payments equal to 6 σ that is 99% of the bell curve and thank you for the help and this is the last payment **$1000.00 !**

What do you think the reaction of the Bank Manager?

29

TYPICAL CORPORATE STRUCTURE

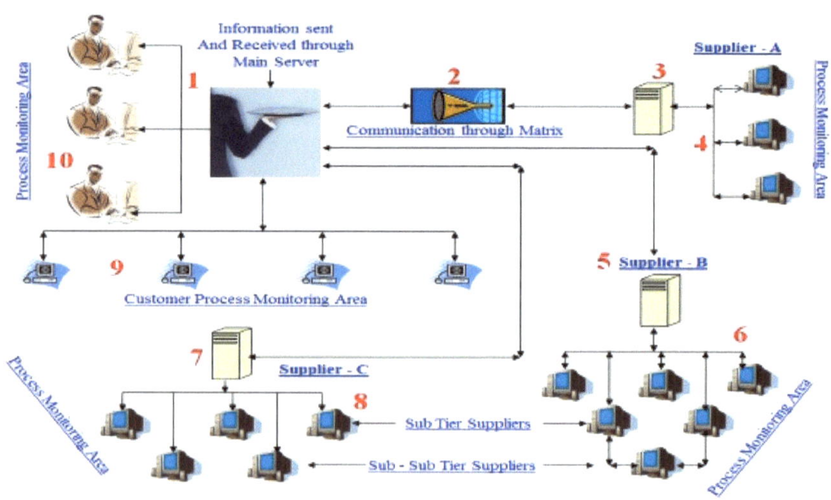

31

www.ingramcontent.com/pod-product-compliance
Lightning Source LLC
La Vergne TN
LVHW051154210926
848486LV00005B/277